James Lawrence Kellogg

A contribution to our knowledge of the morphology of the

lamellibranchiate mollusks

James Lawrence Kellogg

A contribution to our knowledge of the morphology of the lamellibranchiate mollusks

ISBN/EAN: 9783337257422

Printed in Europe, USA, Canada, Australia, Japan

Cover: Foto ©Paul-Georg Meister /pixelio.de

More available books at **www.hansebooks.com**

[ARTICLE 15.—EXTRACTED FROM THE BULLETIN OF THE U. S. FISH COMMISSION
FOR 1890. Pages 389 to 436. Plates LXXIX to XCIV.]

A

CONTRIBUTION TO OUR KNOWLEDGE

OF THE

Morphology of Lamellibranchiate Mollusks.

BY

JAMES L. KELLOGG,

WASHINGTON:
GOVERNMENT PRINTING OFFICE.
1892.

15.—A CONTRIBUTION TO OUR KNOWLEDGE OF THE MORPHOLOGY OF LAMELLIBRANCHIATE MOLLUSKS.*

BY JAMES L. KELLOGG, PH. D.

At the direction of Hon. Marshall McDonald, U. S. Commissioner of Fish and Fisheries, I undertook the following work at the Fish Commission station, Woods Holl, Massachusetts, during the summer of 1891, and while there enjoyed the kindly interest of that gentleman, as well as many attentions shown me by Dr. H. V. Wilson, then biologist in charge. I am also much indebted to Dr. E. A. Andrews, of Johns Hopkins University, for aid and advice. I wish to express my appreciation of the attention and counsel given me by Prof. W. K. Brooks while engaged in this work.

Before I began the work it was pointed out to me by Prof. Brooks, of Johns Hopkins University, that the study of lamellibranch anatomy had been carried on almost entirely by means of dissections, which are difficult to figure or describe satisfactorily, and that comparatively little use had been made of sections. I hope to show not only that the anatomy of a single form may be easily described by this method, but that the comparative anatomy of various forms may be readily demonstrated.

It is interesting to notice, in connection with this use of sections, that the great amount of labor required in producing such a work as Deshayes' Atlas, Mollusques, has been of little service. It is a very large volume of beautiful plates representing dissections; but, even if they had been properly described, the comparison of special organs of different forms would have been very hard to understand. The cost of such a work also renders it inaccessible.

Unfortunately I have been able to obtain but very few representatives of this group of mollusks for examination, and consequently do not feel able to attempt very wide generalizations. Since the completion of the present work a full and valuable paper by Prof. Paul Pelseneer (No. 17) has appeared, dealing principally with forms other than those here discussed. Much has yet to be done in comparative studies in this group, notwithstanding the great works of Lacaze-Duthiers and others.

---

* The principal species of marine bivalve or lamellibranchiate mollusks treated of in the present paper, which the writer has examined personally, are eleven in number, including six forms of greater or less economic value. The structure of other marine species and of the fresh-water mussels *Unio* and *Anodon* is also discussed. The eleven species first referred to are as follows: *Mya arenaria* Linné (common long clam); *Mactra solidissima* Chemnitz (sea clam); *Venus mercenaria* Linné (quahog); *Venericardia borealis* Conrad; *Solenomya velum* Say; *Yoldia limatula* Say; *Arca (Argina) pexata* Gray; *Mytilus edulis* Linné (common mussel); *Pecten irradians* Lamarck (scallop); *Anomia simplex* Verrill; *Ostrea virginiana* Lister (oyster).

A number of elementary anatomical facts are here very briefly mentioned in order to make the explanation of sections illustrating the general anatomy intelligible to those not already well acquainted with it. They are also of use in explaining a number of scattered observations of my own.

THE SHELL.

The outer covering or shell of lamellibranchs is made of an organic base, impregnated by lime, which is taken from the surrounding water. Osborn (No. 14) and others have shown that the mantle lining the shell valves on their inner faces secretes a soft, gummy substance, which soon becomes tough and is finally completely filled with spicules of carbonate of lime. In the oyster, examined by Osborn, no prismatic or mother-of-pearl layer is formed inside this, as is the case in many other forms. The varying proportions of lime make a great difference in the resisting power of the shell. In the shell of *Mytilus edulis*, the common mussel, there is but little lime, compared with the tough, horny basis, and as a result the shell is strong and unyielding. The shell of *Venus mercenaria*, the "little-neck clam" or "quahog," as it is often called on the Atlantic coast, is, on the other hand, made almost entirely of lime. Though the shell is thick and heavy it breaks into many fragments, like a piece of porcelain, when struck a severe blow. The thickness of the shell of many lamellibranchs depends greatly upon the amount of lime held in solution by the water in which it exists.

It is characteristic of this class of the Mollusca that the shell is made of two independent halves, called valves, which are joined to each other by a ligament. This is tough and rubber-like, and between it and the concavity of the shell is the hinge, where the valves, in touching each other, form a fulcrum. In many cases one valve is hollowed out at this point and the other has a corresponding projection, or perhaps many of these, each fitting into a hollow. These relations may be seen in Fig. 96, Pl. XCIV, where *lg* is the ligament and *hg* the hinge. The adductor muscles (*ac* and *pa*), on contracting and closing the shell, cause, through the action of the fulcrum at the hinge, a stretching of the ligament. When these muscles are again relaxed the ligament contracts automatically, as would a piece of rubber, and opens the valves at the opposite extremity.

In the oyster there is no distinct hinge, but the ligament is made of two parts, a central, thick, and elastic portion (Fig. 97, Pl. XCIV, *lg*), and above and below this a slight ridge. The shell projects slightly at these points and may help in functioning as a hinge. The valves of the shell are generally equal to one another in size and shape. In the oyster the left valve is the larger one; it is much heavier and is sufficiently hollowed out to contain the whole of the soft parts (Fig. 97, Pl. XCIV), while the right valve is smaller and almost flat. The animal is attached by the former.

The outer surfaces of the shell are generally marked by concentric lines of growth, and along its edges may be found the horn-like cuticle, secreted by the mantle edge. This cuticular covering of the outer surface of the shell is often thin and may be lost, except at the edges. In the case of *Solenomya* it is very thick, covering the whole shell from sight and extending some distance below its ventral edge. The umbo, a rounded prominence on the dorsal side of many shells, is well marked in *Venus* (Fig. 96, Pl. XCIV).

Internally the valves are white or colored in various tints. In some cases the coloration is due to pigment in the shell substance, in others to the refraction of light by the striated prismatic surface. In the former case, when the color is bright, it may fade at the death of the animal, as I have seen it do in certain species of *Unio*.

The positions of the attachment of the adductor muscles to the shell are indicated by glossy, more or less oval areas, which are sometimes, as in *Ostrea* and *Mytilus*, darkly colored. Very often (*Venus, Mya, Mactra,* etc.) a line of the same glistening appearance runs along near and parallel to the ventral border of the shell, joining at either extremity the adductor muscle scars. This pallial line generally folds inward posteriorly in those forms possessing siphons, making a deep loop into which they may be retracted. The mantle edge is attached to the shell, though not very firmly, throughout the extent of the pallial line.

The shell has sometimes been made the chief or only basis for the classification of the members of the Lamellibranchiata. But a single organ is a very unsafe basis for the comparisons needed in such a case.

ADDUCTOR AND RETRACTOR MUSCLES.

The adductor muscles connect one valve of the shell with the other and are generally very conspicuous. In order to open the shell these must be cut away from their connection with it. Each adductor is generally made of two kinds of fibers, one being of a darker shade than the other. The darker portion of the muscle is generally interior to the lighter, and is in some cases larger, in others smaller, than the latter. The former condition is shown in *Venus* and *Ostrea* (Figs. 96 and 97, Pl. XCIV) and the latter in Fig. 47, Pl. LXXXVI, which represents the adductor of *Pecten irradians*, cut longitudinally and vertically. The darker portion, very much contracted, is shown above the more extensive lighter part. It is well known that, especially in the forms of the *Monomya*, this darker portion of the adductor is made up of fibers presenting a striated appearance, which, however, does not correspond to the cross striation in muscle fibers of arthropods and vertebrates. The lighter part of the adductor is made up of plain, unstriped fibers. Pelseneer (No. 17) confirms the supposition that this condition also exists in many forms with two adductors, which he has examined.

This striation of muscle fibers, which has been referred to by many writers as being present in the adductor fibers of lamellibranchs, has not been described in definite terms, but is generally spoken of as an "appearance of striation." So far as I know, it has only been seen in the adductor muscles. I have found in the muscles of the heart of *Ostrea virginiana* a remarkably distinct striation of the fibers. I have not yet had the time at my disposal for a careful histological examination of the nature of this striation, though it promises to be a favorable subject for such study, as the striations are large and not numerous, there being about twenty to .03 millimeters of the fiber. The muscles of the auricles of this heart at least seem to be composed entirely of these striated fibers.

Fig. 65, a cross-section of a portion of the auricle of the heart in *Ostrea*, will give some idea of the nature of this striation. *Sm* represents an apparently homogeneous basement membrane of the many-layered epithelial wall of the auricle. At certain places it makes a loop out beyond the wall (*a*), perhaps gaining some support to resist, in this way, the contractions of the muscle fibers which are attached to its inner face, though this is very doubtful. These striated fibers (smf) ramify, from their attachment to this membrane, in toward the center of the auricle. They are generally in irregular bundles, though not in the sense that the fibers are closely bound to one another. The fibers are very long.

Scattered among them are many single pigment cells (pgc), which give the auricle its brown color. Also surrounding the fibers are many cells, apparently of an epithelial nature, which seem to be giving off vacuolated ends (r) into irregularly formed spaces, thus having very much the same appearance as the secreting cells of the nephridium. Their nature needs further study.

The functions performed by these two portions of the adductor mentioned above are not well understood. Von Jhering (No. 23) thinks that the part with the plain fibers is used simply as a means of keeping the valves from spreading too far apart on account of the action of the shell ligament. The same view is held by Lankester (No. 8), who does not say, however, which portion he believes to exercise this function. In opposing the view of Von Jhering, Pelseneer (No. 17) says that in the *Pholadidæ*, which have no ligament between the valves of the shell, the adductors are formed entirely of fibers which have no appearance of striation. He thinks that it is probable that, when these two parts exist, "la partie à apparence striée des adducteurs sert, comme chez les autres invertébrés, à produire des contractions rapides."

Although I had made very few observations on the subject, I had come to a different conclusion from any of these. It seems to me that the fibers of the darker portion of the adductor muscles are more compact and firm, and probably supply the greater part of the force required in keeping the shell closed. The fibers of the lighter portion, not being packed so closely together, are able to contract more quickly and close the shell, it may be against a sudden attack. If a valve of *Pecten* be removed, the smaller darker area will be seen to preserve an extremely contracted condition, while the large white part, also partly contracted, now and then makes very sudden and violent contractions, and then immediately relaxes somewhat. These contractions can be made to occur by striking or cutting almost any part of the body besides the adductor, but more especially the mantle edge. The contractions also occur from time to time if the animal is undisturbed. It seems then that a sudden closing of the shell, so often necessary to lamellibranchs, is accomplished by the lighter portion, and that the darker part comes more actively into play when the shell is to be closed for some little time. The lighter portion in *Pecten* is relatively very greatly developed, and, as the very large size of the adductor has been brought about partly for locomotion by its extremely rapid contraction, the lighter part is the one which performs this function. Though other forms do not have this method of locomotion, the manner of contraction of the white and dark fibers may be the same.

The two pairs of foot-retractor muscles are of general occurrence, except in forms with an aborted or absent foot, and are well seen in *Venus*. They are attached to the shell close above the adductors (Fig. 96, Pl. XCIV, afr and pfr), and join the anterior and posterior parts of the foot, respectively. The anterior ones, which pass obliquely backward from their attachment, are shown cut across at ar, Fig. 11, Pl. LXXXI, which is a vertical section in the region of the anterior end of the stomach. In Fig. 12, a little farther back, they are cut more nearly longitudinally and show their final union with one another and the foot, f. The same relations are shown for the posterior retractors in *Venus* in Fig. 17, Pl. LXXXII, at pr.

The muscle system of *Mytilus edulis* is worthy of remark. Fig. 42, Pl. LXXXV, represents the musculature dissected out entire. From the base of the byssus at its anterior portion, two cylindrical muscles, called the anterior foot-retractors, run forward, passing on either side of the mouth (Fig. 35, *ar*), and are attached to the valves of the shell. They are, in the adult, entirely free from any muscles connected with the foot and all their fibers are inserted in the base of the byssus. The tongue-like foot, with its concave anterior surface (Fig. 42, *f*), is continued directly upward to the shell, as two cylindrical muscle bundles (Fig. 42, *fm*), which pass exteriorly to the two anterior muscles just mentioned (Fig. 36, *fm* and *ar*). The foot with these muscles may be removed without disturbing any of the other muscles. These are the posterior foot-retractors. The byssus organ is morphologically a part of the foot, and these muscles described are probably the anterior and posterior foot-retractors, respectively. But the byssus organ has lost all connection with the musculature of the foot, as have the anterior adductors also.

From the base of the byssus, just behind the insertion of the two anterior muscles, extends a great mass of muscle bundles which, attaching themselves to the shell above and posteriorly, serve as its main support. These byssus muscles are arranged in two groups. They are close to one another at the byssus and diverge laterally above to become attached to the valve of either side. (Shown in section in Fig. 37 at *bm*.) The mass extends obliquely backward and is divided with much regularity into a number of large bundles (Fig. 42 *bm*). They are shown in the vertical sections at *bm* in Figs. 38 to 40.

A dorsal muscle, well shown in *Nucula* and *Solenomya*, occupies a position nearly parallel to the anterior foot-retractors (Pelseneer, No. 17, Figs. 7 and 15), and its attachment to the shell is slightly posterior to the latter. In *Pecten* the foot-retractors, and in *Anomia* the byssus muscles, are attached only to the left valve.

Other large muscles are developed in the mantle of many lamellibranchs, in the region of the siphon, which will be spoken of in that connection.

THE FOOT.

This characteristic organ, appearing generally, though sometimes greatly modified, throughout the Mollusca, is in lamellibranchs a muscular projection from the ventral surface of the main body of the animal. It extends more or less anteriorly. In the primitive forms *Yoldia* and *Solenomya*, the lower part of the foot is turned nearly forward, and where the organ has degenerated greatly, as in *Pecten irradians*, it arises from the extreme anterior end of the visceral mass. In *Mya arenaria* it is relatively small, though functional in locomotion and situated far forward. In *Venus mercenaria* and many others, it occupies the whole ventral surface of the visceral mass, extending slightly backward as well as forward, and in a few forms there is a greatly developed heel-like projection posteriorly.

Certain simple and probably primitive lamellibranchs possess a foot which has a more or less circular ventral disc (*Pectunculus, Nuculidæ, Solenomyidæ*). Around the margin of this are a number of thick, short papillae or flutings. Fig. 52, Pl. LXXXVII, representing *Yoldia limatula* with the right valve and mantle fold removed, will show the general relations of the foot (*f*). It is seen to have a volume greater than all the rest of the animal, the mantle being very thin. The ventral disc (*d*) is not expanded (the figure having been sketched from an alcoholic specimen), but its position and the marginal papillæ may be seen. When the foot is contracted, as in the figure, the lateral edges of the disc are brought together ventrally, thus making a crease from before backward through the middle of the disc. The disc may be so fully expanded that the crease sometimes disappears, but this does not often happen.

The most usual form of foot in lamellibranchs is that typically shown by *Venus* (Fig. 96, Pl. XCIV). It is flattened from side to side and extends in this case along nearly the entire ventral surface of the visceral mass (*f*). Its anterior end is plough-share shaped and is greatly protrusible. Instead of being somewhat flat on its ventral surface, it is more or less sharp or heel-like (seen in vertical section at *f*, Fig. 13). In *Mya* the foot is much compressed laterally, and projecting anteriorly to the body, is slightly sharper above than below (in vertical section at *f*, Fig. 23, Pl. LXXXII). It does not extend along the ventral side of the body, or visceral mass, as in *Venus*, but occupies a position more like that of *Mytilus* and *Pecten*.

The foot of *Mytilus* is an entirely muscular, tongue-like organ, flattened dorso-ventrally, concave above and convex on its lower surface (Fig. 42 *f*). In *Pecten* it is relatively much smaller, being a short cylindrical projection from the anterior end of the visceral mass. In *Ostrea* and *Anomia*, which are fixed forms, the foot has entirely disappeared.

The foot serves a number of different purposes, but is generally used as a burrowing organ. The end of the foot is protruded as a long, narrow tongue, which digs into the sand with a worm-like movement, keeping the shell closed as much as possible. When it has penetrated to some depth it expands at the end, the retractor muscles come into play, and the whole animal is gradually pulled beneath the surface of the sand. The forms with the ventral disc are very active. *Yoldia*, when burying itself, makes a sharp point of the anterior part of the folded disc of the foot and very rapidly burrows this into the mud. The disc is now widely expanded, forming a firm anchor, the foot-retractors contract and draw the body down to the end of the foot, in this way quickly covering it. *Solenomya* has the same habit, and also often swims rapidly through the water by using its powerful foot as a paddle. It is stretched out anteriorly, the disc opened, and a rapid backward stroke is made. This is repeated with great rapidity. One or two lamellibranchs have the habit of creeping on the ventral side of the foot, as in gasteropods, but, as far as I have observed it, in the adult of one very small form and in the young of *Mytilus* and *Pecten*, it is done in a very imperfect way, the animal frequently being unable to maintain its erect position when crawling over a smooth surface and falling over on its side.

The foot of many lamellibranchs, as that of *Venus* or *Anodon*, is made up of muscle fibers, which are irregularly distributed vertically and horizontally, leaving everywhere spaces which are in connection with the vascular system. Blood being forced into these spaces by the heart, causes the extension of the foot. The general direction of the fibers in the foot is indicated in Fig. 13, Pl. LXXXI, *f*. In this section a sharp separation occurs between the foot and the genital gland above (*g*). Just below this the foot shows many transverse fibers (*ms*). From this region, too, three principal bands of muscle fibers extend down toward the ventral side of the foot. Their contraction, probably, aids the foot-retractors in drawing the foot up close to the visceral mass. Farther back (Fig. 14), in the region of the posterior end of the stomach, the sexual gland forces its way for a considerable distance down between the more scattering foot muscles. This very generally occurs in forms with this kind of foot. In *Yoldia* and *Solenomya*, also, the sexual gland occupies a considerable portion of the upper part of the foot. The walls of the foot are made of a more dense layer of muscle fibers (Fig. 13).

In such forms as *Mytilus*, where the foot is degenerated and is of little or no use to the adult animal as a burrowing organ, it is not necessary that it should be expanded, and it has almost entirely lost its blood spaces, those only remaining which may contain blood for the nourishment of the tissues of the foot. The fibers are closely packed together, making the foot very dense and tough.

THE BYSSUS.

This organ is generally considered to be a gland of the foot. The byssus itself is made of a number of horny secreted fibers which attach by their outer ends to foreign objects. The part which does the secreting, the byssus organ, occupies various positions in different lamellibranchs. In *Nucula* there is a small blind sac near the posterior edge of the disc of the foot, and Pelseneer has described above this several gland-like cells which he thinks represent the byssus organ. In *Venericardia borealis* a well-developed byssus is present in a slight groove near the middle of the ventral surface of the foot. In *Mytilus* the great byssus has no connection with the foot in the adult, but is situated behind it.

The byssus organ of *Venericardia* is one of the best for examination, as it is not greatly complicated. Fig. 73, Pl. LXXXIX. represents a horizontal section through it. The secreting surface is deeply folded (*fd*), and in these folds the secretion is seen in long sheets (*bs*). Surrounding the folds is a mass of the secretion made of concentric layers, which have been added to its inner border. At the inner ends of the folds are many vertical muscles (*bm*) which are strongly inserted and serve by their attachment to the valves of the shell above as a very powerful support. Among these muscles are many large, almost clear cells (*c*).

Fig. 74 shows the epithelial surface of one of these folds. The greater number of the lining cells approach to a columnar form (*cc*), and appear to be ciliated. At the deeper part of the fold the lining cells suddenly become very large, indistinct (*lc*), and almost entirely unstained. I could make out no nuclei in them. The cells which seem from the appearance of the section to do the secreting, are the epithelial cells (*cc*). The byssus secretion (*bs*) I never found extending down over the large clear cells, but it more often adhered to the other cells as shown in the figure, and was much thicker at the outer than at the inner end. In some sections the dense apparent ciliation of these cells suggested that possibly there was really a striated secretion instead of a ciliation.

If individuals of *Mytilus*, which have been torn from their attachment, are put in a dish of sea water, they soon become again attached.* A very fine, transparent thread appears, and where it strikes a solid body, its end spreads out in a number of root-like processes which form a firm attachment. Soon another similar thread appears and attaches itself, generally at some distance from the first. Though this may take place in a minute or two after the individual is placed in the dish, I was not able to see the exact manner in which the thread was protruded. I imagine, however, that

* Since the above was written, I have carefully observed the method of attachment in *Mytilus*, which is accomplished by the foot, as referred to by Prof. Verrill. An account of it will be included in a later paper.

at the will of the animal a fine stream of the secretion was thrown out and that it soon hardened after coming in contact with the water. The young individuals are able, in some way, to leave their attached byssus, wander about by means of the foot, and reattach themselves in new localities.

THE MANTLE.

If a lamellibranch be taken out of the shell, the whole animal will be seen to be covered by two fleshy flaps or folds (*m* in the figures) generally united to each other in the middorsal line and attached to the top or the sides of the visceral mass and to the adductor muscles which pierce them to become attached to the shell. Ventrally these folds hang down and cover the gills and foot.

In many cases, as in *Nucula, Yoldia, Arca, Trigonia, Pecten,* and *Anomia,* the ventral edges are free from each other and are not in concrescence with the gills. In the oyster (*Ostrea virginiana*) the mantle folds are not connected with each other ventrally, but are connected with the outer lamella of the outer gills. In other lamellibranchs the ventral borders are fused and are connected with the gills. *Solenomya* is an exception to this class in that the mantle is fused and is not connected with the gills.

Mya (Figs. 23 to 31, inclusive) is a good example of the ventral concrescence of the mantle. In such cases there are left two small posterior apertures, so that water may pass in and out of the mantle chamber, and a larger anterior one for the protrusion of the foot. These three openings are always present in such cases, except in *Solenomya.* Here there are but two openings, a large anterior one for the foot and a single posterior opening for both exhalent and inhalent streams of water. This has also been noticed by Pelseneer, but he has not spoken of the method employed in separating the exhalent and inhalent streams. If *Solenomya* be put in an aquarium it gradually opens the valves of the shell, and the posterior opening may be seen to have the appearance of a single slit, as represented in Fig. 61A, Pl. LXXXVII, whose edges bear a number of tentacles. Often the sides of the slit approach each other in the center, the upper and lower ends assume a circular shape, and there are formed a lower inhalent and an upper exhalent opening (Fig. 61B). This position is quite constantly kept as long as the animal is undisturbed, and is similar to the condition of *Anodon,* where there is no actual concrescence between the two openings. In *Anodon,* however, the mantle edges ventrally are nowhere fused.

In *Mytilus* (Figs. 33 to 41), the mantle edges lie very close to each other, but are not actually united.

As has been stated, the mantle is fused to the visceral mass above or on the sides. In the oyster (*Ostrea virginiana*), the folds are not thus connected with the visceral mass at all points on both sides. They are in concrescence on the left side, which lies deep in the hollow of the fixed valve, but the right side is modified. Over the pericardial cavity and that portion of the visceral mass immediately anterior to it, the mantle is perfectly free from its dorsal border as far ventrally as the concrescence between it and the outer lamella of the outer gill. Back of the pericardium the mantle is again connected to the anterior border of the adductor. A very peculiar cavity is thus formed, on the right side only, and chiefly over the pericardium. It opens dorsally to the exterior, and its lower border opens into the epibranchial chamber, which, in this region, receives water from both gills.

This chamber may be seen in section at *epe* in Fig. 5, Plate LXXX. where the epibranchial chamber is extended upward past the visceral mass, pericardium, and rectum, and to the exterior dorsally. In the upper part of the figure the mantle is represented as closely applied to the body, but in the live animal it is widely opened. Fig. 3, just posterior to the stomach, shows the union of the mantle and visceral mass near the anterior border of this cavity.

It is probable that some of the water coming into the epibranchial chamber through the gill lamellæ passes again to the exterior by means of this unusual path, instead of going posteriorly through the epibranchial and cloacal chambers. This may be the case, because the mantle is here loose and not applied closely to the body, and the channel thus afforded is directly in the line of the currents from the gill into the epibranchial chamber. It probably in no way aids in the aëration of blood by bathing so much of the mantle wall with water, for the latter is not richly supplied with blood spaces, as in forms like *Venus* and *Anodon*. What may have caused this asymmetrical condition to appear I am unable to conjecture.

The mantle edge is more exposed than any other part of the body between the valves of the shell, excepting the foot when it is extended. The protrusible foot may be closely contracted to the visceral mass, but the mantle edge, though it may be drawn away from the edge of the shell, is always capable of less retraction. Fig. 96, Pl. XCIV. shows the maximum of contraction of foot and mantle in *Venus*, and Fig. 97, the great degree to which the mantle of the oyster (*mc*) may be contracted. On account of its close contact with the exterior, the mantle edge in all forms is relatively greatly thickened, and in it have been developed sensory organs, those of touch and in a few instances of vision.

This muscular mantle edge generally possesses three primary longitudinal folds seen in section in many of the figures at *me*. In some cases, a primary fold may become greatly enlarged and broken up into several secondary folds. The folds are least marked anteriorly. They are generally pigmented, and most deeply in the ventral and posterior extent of the mantle edge. The cells of certain of these folds secrete the horny cuticle, which is reflected outwardly over the growing edge of the shell (Fig. 23, *c*). Eyes and tentacles are frequently—the latter always—present in definite folds (Patten, No. 15; Rawitz, No. 21). Over the outer surface of the mantle, next to the shell, are many gland cells, which secrete a sticky substance that becomes impregnated with lime and forms new shell layers.

THE SIPHONAL REGION OF THE MANTLE.

In lamellibranchs the posterior parts of the mantle lobes are variously modified to form separate openings for the inflow and outflow of water. In certain cases, where the two mantle folds are free from one another, though generally opposed throughout most of their length, they spread apart posteriorly in two regions close together, making a lower inhalent and an upper exhalent opening. This is very conspicuously shown in *Unio* or *Anodon*. A similar arrangement has already been described for *Solenomya*. These openings are guarded by greatly developed tentacles, which are also generally present at the ends of the siphons. Prof. Brooks has described (No. 2) an enormously developed unpaired tentacle situated on the mantle edge of *Yoldia*, on the right side, and near the base of the siphons. This sense organ, supplied with an axial nerve, may be extended out beyond the ends of the siphons.

In the oyster the partition between the two openings in the mantle is permanent, the mantle folds being united. Fig. 89, Pl. XCIII, represents a view of this part of the mantle seen from behind. The mantle folds are fused at *br p*. Above, the upper part of the cloacal chamber is exposed where it communicates with the exterior, and the rectum (*r*) is seen to open into it. Below the fusion, the branchial chamber is seen, together with the posterior ends of the gills (*og* and *ig*), which hang in it.

In *Mytilus* this region is more modified. The cloacal opening, which is small, appears from the exterior to be much more definite (Fig. 87, Pl. XCIII, *co*) and is surmounted by a tough ring, which, like the whole mantle edge in this region, is colored by a deep-brown pigment. Tentacles are here absent from the mantle edge. Ventral to this, the mantle folds do not shut in a complete branchial orifice. Closing the upper part of the space between the mantle folds on the inside, is a thick, pigmented membrane, a part of the mantle (*br m*). This is also shown in Fig. 88, which is a posterior view of Fig. 87. In this figure the extreme posterior ends of the gills mark the partition between the epibranchial chamber above, whose opening to the exterior is seen at *co*, and the branchial chamber below. Thus far there is little indication of a development of much tissue in connection with the siphon-like openings other than that usually present in the mantle edge.

If we imagine that the two separated openings in *Ostrea* and *Mytilus* are made definite—that is, that the branchial passage is separated from the ventral mantle edge by a second fusion—and that they are protruded as tubes, we will have in the main the condition in the siphoned forms (Fig. 96, Pl. XCIV, *sn*). A vertical, longitudinal section through the posterior region of the body of *Venus* (Fig. 93, Pl. XCIII) will show the relations of most of these parts. Below, and to the right in the figure, is seen the branchial chamber (*br c*), in which hang the gills (*ig*). The upper part of the gills, forming the floor of the epibranchial chamber (*ep c*), is seen in the section, and the openings of the water tubes into the latter chamber are indicated. Bounding this part of the epibranchial chamber above is the posterior adductor (*pa*). The epibranchial chamber opens into the base of the cloacal siphon, which also receives the end of the rectum (*r*) from above, it having come down over the adductor. This basal portion of the siphon may perhaps be called the cloaca, though it is small. The lower or branchial siphon opens into the branchial chamber, but at its base there stretches across its whole upper part a membrane (*br m*), occupying the same position as that already referred to in *Mytilus*. If we should take an unsectioned specimen and throw back the mouth folds so as to get a view of the base of the lower siphon from the branchial chamber, we would have the membrane shown as in Fig. 90, Pl. XCIII, *br m*. Extending down from the posterior end of the gills, it covers all but the lower part of the base of the branchial siphon seen below in the figure. It does not extend straight across from one side to the other, but presents the appearance of a deep notch extending upward. On either side of it the bases of the thick siphon walls may be seen.

One might naturally suppose that from its position this branchial membrane would allow water to enter the branchial chamber, but that any back flow would apply it over the base of the branchial siphon; this would prevent the water from escaping, thus acting as a valve. If the animal is able or has occasion to use this fold in such a way, I have not observed it. *Mactra solidissima*, the sea clam, much like *Venus* in anatomical points, has a branchial membrane that apparently completely covers the branchial siphon at its base. But if an individual be quickly taken out of the water the valves close and the water in the branchial chamber is discharged principally through the *branchial* siphon, so that the membrane must have been raised in order that this might occur.

If the valves of the shell should suddenly close, much of the water in the branchial chamber would escape ventrally. While this is going on, the siphons are being drawn within the shell, and of necessity against this pressure in the branchial chamber, caused by the closing of the shell. If, instead of being closed, the branchial siphon should be opened, and thus allow the imprisoned water to rush out, it would allow their retraction to be more easily and quickly accomplished.

The siphons of *Venus*, however, are small, and do not meet the difficulty in retraction encountered by the enormously developed siphons of *Mya arenaria*, the "long-necked clam," especially as the branchial chamber is here closed below over nearly its entire extent, and allows little water to escape between the mantle edges. When the siphons are contracted—and the process is always comparatively a very slow one—a stream of water is discharged from both, but mainly from the branchial. Though finally brought within the shell, their outer ends are somewhat exposed, as the shell in their region is expanded and its valves can not meet behind them. The branchial membrane is not here present, but may perhaps be represented by part of the thickening in the partition between the siphons (Fig. 94, Pl. XCIII, *br m*). If this is so, the organ may have been lost because of its interference, though it may have been slight, in the laborious process of withdrawing the siphons into the shell.

Just what advantage may be subserved in the forms where the branchial membrane is so greatly developed is not apparent to me.

The series of figures 18 to 22 represents this posterior region of the body of *Venus* in vertical section. Fig. 18 has been cut just in front of the branchial membrane (*br m*) and the posterior adductor is seen above. The mantle edge has become muscular and very thick at *m*, to form, farther back, the walls of the siphons. The bases of these have been cut across in Fig. 19, and it may be here seen how their walls are gradually constricted off from the mantle at *x*. A fold of the mantle (*m*) extends across under the lower siphon and is also present in Fig. 20. In this latter figure the mantle fold is entirely separated from the siphon walls, except dorsally, and these walls are seen to be very thick and muscular, especially those of the cloacal siphon. In Fig. 21 they have assumed a uniform thickness. In Fig. 22 the siphons are cut across where they have protruded backward beyond the mantle edge. The basal part of the lumen of the upper or cloacal siphon (Fig. 19, *ns*) is somewhat triangular in section, while that of the branchial siphon is more nearly circular. Toward their outer end they appear as slits elongated dorso-ventrally, though not to so great an extent in the living animal.

The anatomy of the siphons of *Mya* is much the same as in *Venus*. Fig. 29 is a thick section just before the bases of the siphonal openings. The cloacal chamber is cut across at *cl*, showing the gills at their posterior ends, separating the cloacal from the branchial chamber. The posterior end of the partition between the siphons is seen at *br m*. The muscles, which are to become the siphon walls farther back, are shown at *ns*. These, from this region to the ends of the siphons, are covered by a thick, gelatinous, semitransparent tissue, *et*. Still farther back, as in Fig. 30, in which section the right side has been cut deeper than the left, are seen the siphon walls on all sides, the whole still within the mantle folds. Fig. 31 represents a section of the siphons at some distance from the body.

The muscles of the siphon of *Yoldia* may be divided into two chief groups: (*a*) transverse and (*b*) longitudinal layers. As far as I know, there are no circular fibers. Small transverse and longitudinal layers alternate with one another to form the siphon walls. Fig. 53, Pl. LXXXVII, represents a transverse section of a bit of the siphon walls in this form, *as* being the cavity of the anal and *bs* the branchial siphon; *ss* is the septum separating them. The transverse layers (*trm*) extend across the walls. Nuclei are present most frequently in the central, narrower portion. At the outer edges of the layers the fibers separate and spread out to become attached. In this region, also, numerous nuclei of the fibers appear. The longitudinal muscles occur principally between the transverse layers (*lm*), but there are also many smaller bundles near the point of attachment of the transverse muscles. The whole siphon wall outside and inside is covered by a jelly-like tissue, containing nuclei, and at places showing elongated cell-walls, *c*.

THE DIGESTIVE TRACT.

The mouth.—The mouth is situated in the median line between the two labial palps, and just behind the anterior adductor muscle when it is present. It is not sharply marked off from the œsophagus, being a funnel-shaped opening for it.

The palps.—These are two lips or folds anterior and posterior to the mouth. They extend backward on either side toward the anterior ends of the gills. They are often large and plate-like (*Nucula* and *Yoldia*). In other cases they may be prolonged as narrow bands (*Mytilus*) or they may be short and thick (*Ostrea*). The first case is shown in Fig. 52, Pl. LXXXVII, at *p*, where the palp extends from the mouth back to the gills (*g*) at the posterior end of the visceral mass. Here and in *Nucula* (Mitsukuri, No. 13) the palps possess a long appendage (*ap*) supposed to aid in the collection of food. These may be protruded to the exterior just below the siphons.

Fig. 95, Pl. XCIII, shows the relation of the palps to the mouth, as well as their general shape. The figure represents the anterior end of the body of *Mytilus*, cut in a vertical transverse plane just posterior to the mouth. The most posterior, or inner palp (*ip*), extending on either side of the mouth opening, hides the latter, which is situated just above *mo*. The outer palp (*op*) occupies a similar position before the mouth. A dorso-ventral striation is seen to exist over the lower three-fourths of the inner surface of the palp (at *op*), and the surface of the inner palp, opposed to it, is similarly thrown into folds or ridges.

These ridges have very much the same appearance to the unaided eye as the gill filaments, and have led many observers to mistake the palps for gills. They are simply ciliated ridges occurring on but one side of each palp—that next to the mouth.

Sections across these folds would differ somewhat in appearance according to the regions from which they were taken. Fig. 62, Pl. LXXXVII, represents a section across the folds of the palp of *Ostrea* in the ventral region, in this case farthest away from their attached basal portion (Fig. 97, Pl. XCIV. *ip*). The folds at this free edge of the palp are generally thicker than at the base, and are thrown into two or three small secondary folds (*sf*). The ciliated epithelium covering the folds is composed of much elongated cells, more greatly developed on the side of the fold on which occur the secondary folds. The bases of the folds rest upon a more or less complete connective tissue layer (ct). The transparent, irregular cells composing the main body of the palp, and extending into the folds, are the same as those found in the mouth and walls of the visceral mass of Ostrea, and described by Prof. Brooks (No. 3) as fat cells.

Fig. 63 is taken from about the middle of the palp. Here the folds are more narrow, and the irregular secondary folds on the side of the primary folds have given way to one regular secondary fold (sf). In Fig. 64, a section close to the attached border of the palp, the folds are very regular in size, and without secondary ridges.

The palps of Pecten are peculiar in having upon their free edges near the mouth a number of projections which are extremely convoluted and give the appearance of a heavy fringe, part of which is indicated in the palp just anterior to the mouth at fr, Fig. 43, Pl. LXXXV. Fig. 68, Pl. LXXXVIII, represents a small portion to show the nature of this fringe. The small figure at the right (A) represents a large branch, whose base is continuous with the free edge of the palp near the mouth. The trunk of this tree-like mass is not solid, but is merely a thin sheet of palp tissue whose surface nearest the mouth is concave. The outer surface shown in the figure is convex. A section across it would be somewhat crescent-shaped. Fig. 68 (B) represents a part of the extreme tip of the fringe more highly magnified. Here the concave side of the mass is presented, and it shows that the whole fringe is made by flat, sheet-like outgrowths of the edge of the palp. The edges of these flat projections turn inward toward the mouth. These edges seem to have grown a great deal more rapidly than the interior of the sheet, and have thus been forced to convolute themselves greatly, as represented in the figure.

A section through the ends of this fringe shows an epithelium made up of very much elongated, ciliated cells (Fig. 75, Pl. XC, ep), whose nuclei are arranged in quite a definite row in their outer third. At many points in this epithelium are found groups of large gland cells (gle). At the base of the cells is a more or less definite basement membrane (bm). Running through the compact tissue of the fringe are blood vessels whose flat, bounding endothelium is plainly seen (br).

The labial palps take food collected upon the gills from the anterior ends of the latter, and by the cilia pass it on into the mouth. The movement of food particles here does not seem so rapid as upon the gills, and its path seems much less defined than the latter. This is easily seen upon the palps of Yoldia. The long, ciliated palp appendage of this form with its convoluted borders is similar to the fringe about the mouth of Pecten, though situated at the posterior end of the palp. The appendage in Nucula, like that of Yoldia, is supposed by Mitsukuri (No. 13) to serve in collecting food, and the mouth fringe of Pecten in all probability has a similar function, the products of its gland cells cementing the food particles together as on the gills, and its ciliated epithelium passing this on down its concave inner surfaces to the mouth.

The œsophagus.—In the Nuculidæ, Pelseneer (No. 17) has described in the buccal region of the digestive tract a transversely enlarged glandular portion, called the pharyngeal cavity. It occurs in no other lamellibranch, but he thinks is homologous with a cavity found in other mollusks—Patella, Fissurella, Haliotis, and Dentalium.

The œsophagus proceeds nearly vertically upward to the stomach. It is very large in Pecten (Fig. 43, mo), smaller in Mytilus (Fig. 33) and Venus (Fig. 10). Its opening into the stomach is generally somewhat funnel-shaped, but in Mya is abrupt, with a definite muscular opening.

The stomach.—This organ is a greater or less enlargement of the alimentary canal, and is placed in the dorsal part of the visceral mass, its long axis generally being that of the body. In *Yoldia*, however, it is dorso-ventral (Fig. 69). In many cases, as in *Venus*, it is close to the dorsal wall of the visceral mass (Fig. 12, *s*). Frequently in individual cases it lies much more in one side of the visceral mass than in the other. The walls are often irregular (Figs. 12, 26, 44, *s*). Closely applied to the stomach walls is the liver mass, whose ducts open into the stomach at different points. The number of these openings varies in different forms.

The intestine.—The stomach generally narrows posteriorly to open into the intestine (*i* in the figures). In most cases this is coiled. In the degenerate deep-sea forms recently studied by Pelseneer (No. 17), this observer describes the intestine as being so short as to be almost straight. He supposes that this condition has been brought about by the carnivorous habit of the animals, which is inferred from the animal matter found in the digestive tract.

The intestine is but little coiled in *Yoldia*. Fig. 69, Pl. LXXXVIII, represents the whole tract in this form. The mouth (*m*), short œsophagus, and dorso-ventrally elongated stomach (*s*), are seen anteriorly. From the bottom of the stomach springs the intestine (*i*). This proceeds backward, then upward as far as the top of the stomach. With a bend forward, it makes a loop, always on the right side of the stomach, before running backward over the posterior adductor (*pa*) to empty as the rectum (*r*) into the cloacal siphon.

In other lamellibranchs the intestine is generally much more convoluted. In the oyster, for example, it runs downward and backward from the stomach so far as to be ventral to the adductor muscle. It returns to the upper part of the stomach, makes a complete loop around it, and then proceeds back over the pericardium and adductor muscle. The extreme end of the rectum hangs close to the base of the cloacal siphon in forms where it is present (Figs. 93, 94, *r*); in other cases it projects well into the cloacal chamber (Figs 89, 97, *r*). The extreme end is often slightly thickened.

Crystalline style.—This organ is a diverticulum of the stomach or the intestine, generally close to the latter. In the diverticulum, where it is much developed, is the transparent style, evidently a product of the secretion of the epithelium of the diverticulum. Pelseneer says that the stomachs of all lamellibranchs are lined with a cuticular covering which is continuous with the style when the latter is present.

Fig. 26, Pl. LXXXIII, represents the greatly developed style of *Mya arenaria* (*cst*) arising from the bottom of the posterior end of the stomach (*s*) and running ventrally with a bend to the right, so nearly in a vertical plane as to be shown all the way in the section. Reaching the ventral part of the visceral mass, it runs forward a short distance (Fig. 25, *cst*) and ends blindly. In *Anomia* it even runs out of the visceral mass into the mantle edge.

The style may be easily drawn out of the sheath-like diverticulum. In cross-section, it appears to consist of innumerable concentric lamellæ. If the large style of *Mactra* be taken out of the animal and an end of it carefully teased, we find that it seems to have a central, apparently softer axis, around which the lamellæ are deposited in a concentric spiral. When an end of any width is artificially made, as represented in Fig. 55, Pl. LXXXVII, this outer portion may be unwound to any extent from the central axis, showing its spiral arrangement.

The crystalline style appears here and there in various groups. It may be present in one form and entirely absent in another closely related to it. In the primitive *Nuculidæ* it is represented by a mere rudiment (Pelseneer). It has been homologized with the radular sac of the *Glossophora*, but probably not correctly, on account of its point of origin. Its function also is unknown. It has been regarded as a store of reserve food material. Barrois (No. 1) and Pelseneer suppose that its purpose is to surround sharp particles in the digestive tract, which might injure its lining epithelium. Such a function seems to me improbable. It is generally supposed that food is taken into the mouth and stomach by ciliary action only. In many forms large quantities of sand are taken in by the same means. It would be impossible for the style substance to protect the stomach walls from such a mass of foreign bodies by covering them. Only when an extraordinarily large and sharp piece enters, could this function of protecting the stomach take place, which seems altogether improbable. The lining cilia of mouth and œsophagus could probably not pass into the stomach a foreign body much larger than a grain of sand. The digestive tracts of those forms which have no style are probably not easily injured.

THE LIVER.

This gland is paired, there being one-half on each side of the visceral mass. In cutting into the visceral mass, however, the dark-brown gland surrounding the stomach gives no appearance of being of two parts. If the stomach be injected, it will be found that the injecting substance has penetrated the liver mass through its ducts, which open into the stomach. The ducts are, in the main, very fine and traverse the liver in every direction.

The openings of the ducts into the stomach are usually large and cause much irregularity in its walls (Fig. 2, Pl. LXXIX, *s*). The position of this mass and its extent in *Ostrea* are shown in the figure to which reference has just been made (*l*). It extends in this region of the stomach from near the mouth dorsally, to the extreme ventral wall of the visceral mass. It nowhere touches the walls of the visceral mass excepting below, being surrounded by the sexual gland (*g*). Its extent in *Venus* is readily seen by referring to Figs. 11, 12, and 13, *l*, vertical sections approximately through the anterior, middle, and posterior parts of the stomach. In this case, in the oyster, in *Cardita* (Fig. 51, *l*), and in *Mytilus* (Figs. 33 and 34), the genital gland (*g*) more or less completely surrounds the liver. In *Pecten* (Fig. 44. *l*) the liver in the region of the stomach is only bounded by the sexual gland on its ventral surface. This is also the case in *Mya*, shown in Fig. 25. *l*. The posterior end of the stomach in this form (Fig. 26) is not surrounded by the liver, but by the sexual gland. Though varying a great deal in size in different lamellibranchs, the liver seldom, if ever, extends farther backward than the posterior end of the stomach. It extends forward, however, to the anterior end of the visceral mass, as is shown in the sections anterior to the stomach. The posterior end of the liver is irregular, and in vertical sections of this region portions of the sexual gland may be seen surrounded by the liver mass (Fig. 34, *g*). The boundaries of both glands are irregular and they everywhere lie closely applied to one another.

As a general thing, the secreting tubules of the liver are packed together very closely. In the more primitive forms, however, as *Nucula*, *Yoldia*, and *Solenomya*, the liver mass is not at all compact. In *Yoldia* the liver tubules, surrounded by the sexual glard, are free from one another throughout most of the extent of the gland, being only connected by long branching threads of a connective tissue.

THE GENERATIVE MASS.

The greater part of the visceral mass is made up of this organ and the liver. It is of a much lighter color than the latter gland and varies much in its outward appearance in different forms. It is a large gland, surrounding the liver, and constitutes the posterior part of the visceral mass. It grows in between bundles of muscle fibers of the foot, wherever spaces may be left. In *Ostrea* it reaches the body wall anteriorly, surrounds the liver, and closely invests the intestine throughout almost its entire extent. A part of the visceral mass extends down under the pericardium, and backward for some distance below the adductor, where it ends in a blunt, rounded point, forming most of the anterior boundary of the cloacal chamber. The intestine runs almost to this extreme tip and then returns toward the stomach. It is, in this region, under the adductor muscle, entirely surrounded by the sexual gland (Fig. 7, Pl. LXXX, *g*), which is also seen at the extreme posterior part of the body represented in Fig. 8, *g*. In the oyster, also, the visceral mass extends in much the same way above the pericardial cavity and over part of the adductor. It consists here of the rectal part of the intestine, surrounded by a layer of the generative gland, which becomes thinner and thinner until it finally disappears, and the extreme tip of the rectum is continued on into the cloaca without this covering. This is seen in section above the heart region at *g*, Figs. 4, 5, 6, and 7. In Fig. 6 it becomes thinner, and in Fig. 7 posterior to the pericardial chamber, and over the adductor it has almost disappeared. This backward extension surrounding the rectum is an unusual one. In *Venus* the generative gland penetrates into spaces between the uppermost muscle bundles of the foot, as is usual in forms with a locomotor foot. The posterior part of the visceral mass has many scattered muscle bundles, generally transverse, as indicated in Figs. 14 and 15, *mf*, running from one side to the other. The sexual gland pushes down among these muscles for a considerable distance. In a case like *Yoldia*, where the organs of the visceral mass are not at all crowded, the sexual gland still occupies a considerable part of the base of the foot. The definite boundaries of the gland in *Mya* will be seen by a glance at Figs. 24 to 26, Pl. LXXXIII, *g*.

In *Anomia*, greatly modified in many ways on account of its fixed condition, the sexual gland is very asymmetrical, extending out into the mantle on the right side. In *Mytilus* the foot (not present in *Anomia*) is entirely muscular and contains none of the sexual gland. Much of the visceral mass, also, is occupied by the large byssus muscles, and, as a result, the generative gland has pushed out into the mantle lobes of both sides and completely fills them, as seen in Figs. 32 to 41, *m*.

The ducts leading from the sexual glands open in a variety of ways. They may open directly to the exterior or into the excretory organs. If into the latter, they may open near its pericardial end, its middle, or its external end. It is probable, from the fact that in the most primitive forms this gland opens into the pericardial end of the kidney, that the free opening to the exterior is, in some cases, a secondary condition, as shown by Pelseneer (No. 17).

In a majority of cases the sexes are separate, but hermaphroditism exists in very many forms. Pelseneer points out the fact that it exists in isolated forms, species in certain genera, as in *Ostrea*, *Pecten*, and *Cardium*, and whole genera in certain families, as in the case of *Cyclus*, *Pisidium*, and *Entovalva*. He says: "Chez tous ces animaux la glande génitale *elle-même* est hermaphrodite; elle produit dans toute son étendue des œufs et des spermatozoïdes, ou bien si une partie est spécialisée pour donner naissance à chacun de ces produits, ceux-ci sont amenés au dehors par une seul canal, *hermaphrodite (Pecten)*."

The European oyster, *Ostrea edulis*, is hermaphrodite, but in the American form, *O. virginiana*, the sexes are separate. While rearing the young of this form from the eggs at Woods Holl, with Mr. Harrison of the Johns Hopkins University, we found a specimen apparently containing both eggs and spermatozoa. On sectioning parts of the generative gland, I found it to be hermaphrodite, as was suspected. The large follicles (Fig. 72, Pl. LXXXIX) were generally more or less united, and their lumens everywhere, in specimens taken from a number of different parts of the glands, appeared full of ripe spermatozoa. The ova, with their distinct nuclei, were apparently unripe, though many were free from the follicle walls. The majority of them were yet attached. The cells of the follicular epithelium showed ova in all stages of growth, and none of them, apparently, were giving rise to spermatozoa.

This specimen was obtained late in June, near the end of the breeding season. Whether or not its sexual glands indicate the change in function from male to female or from female to male, which possibility has been suggested for some lamellibranchs, I can not say; but, as abnormal hermaphroditism often occurs in all groups of animals, this may be an example of it.

The sexual gland of *Pecten irradians* is hermaphrodite, and there are an ovary and a testis on each side of the visceral mass. Both glands are ventral to the liver (Fig. 44, *g*) and have a spongy appearance. The testis is the more ventral of the two and is of a cream color. The ovary, situated above this, has a reddish hue, which is very marked in the living animal.

Fig. 71 represents a section passing vertically through the outer wall of the visceral mass, where the testis and ovary are closely apposed. The body wall is represented at *ep* and consists of a single layer of columnar, ciliated, epithelium cells, whose nuclei are about equally distant from their outer ends and the thick basement membrane (*bm*). In this epithelium are many conspicuous gland cells (*glc*). Between it and the follicles of the generative gland is a thick layer of connective tissue, extending in between the follicles. The follicles of the ovary (*or*) are not so regular in outline when seen in section as those of the testis (*t*). The walls of the latter bear a follicular epithelium (*fep*). In the ovary, the cells of this layer are in all stages of development into eggs. The eggs themselves, crowding the follicles, possess a very thick egg membrane and their protoplasm is finely granular. A duct from the follicles is seen at *d*.

The mother cells of the spermatozoa (*fep*) are circular and of constant size in the follicles of the testis (*t*). As we follow the mass of cells inward from these mother cells they become very gradually smaller and smaller, until their final divisions result in the spermatozoa. These are so arranged that their "tails," in forming, project in extended masses toward the lumen of the follicle and give it a radiating appearance.

I have not been able to determine how many times a mother cell divides in forming spermatozoa, for the cells are all rounded and give no evidence of their divisions, as they do in the testes of many animals. A duct of the testis containing spermatozoa is shown at d. The ducts of both testis and ovary are composed of slightly columnar, ciliated cells. In the wall of the duct of the testis is shown a single deeply stained cell, which is evidently a gland cell.

In the mother sperm cells of *Mytilus* the chromatin is arranged in a crescent-shaped mass at one side of the nucleus.

The spermatozoa vary greatly in different forms, both in size and general shape. In a single follicle of the gland of any individual, also, they are of various shapes, often very markedly different from the normal.

Fig. 67, Pl. LXXXVIII, represents a very few spermatozoa illustrating these points. The "tails" in all these cases are more than twice the length represented in the figure. The sharply pointed spermatozoan "head" of *Yoldia* is shown (a) to be similar to that of *Venus*, (b) in that both are elongated and conical in outline. The latter, however, are always bent. In a very large number of cases the tail proceeds from the narrower end of the head (c). The spermatozoön of *Pecten* is represented at d. It is actually much shorter than that of *Venus*. They frequently show a form similar to f. The spermatozoön of the oyster (e) has a nearly spherical head, which gradually tapers off into the tail. If spermatozoa are characteristic even of species, as has been suggested, it may be of interest that those of *Anomia* are very similar to those of the oyster, since Pelseneer does not regard these forms as being in any way closely related.

THE VASCULAR SYSTEM.

The heart, the chief organ of the circulatory system, is in the majority of cases situated on the dorsal part of the body and is greatly elongated from before backward. It consists of a muscular ventricle and two lateral, generally more delicate, auricles opening into it, one on either side. The organ lies in an extensive pericardium. The usual relations in position between these parts may be seen in Fig. 45, Pl. LXXXVI, where *ven* represents the ventricle, *au* the auricle, and *p* the pericardium. In a few cases, as in the oyster, the heart has changed from its usual position, and its long axis is dorso-ventral. In Fig. 97, Pl. XCIV, the organ may be seen represented as lying beneath the pericardium, just in front of the adductor muscle. The ventricle is most dorsal, and the two auricles open into it ventrally; these are pigmented; they receive blood from the gills.

As a rule the ventricle of the heart is traversed by the rectum, but this does not occur in *Nucula, Area, Anomia, Meleagrina, Ostrea,* or *Teredo*. Its position in the ventricle of *Pecten* is shown in Fig. 45, r. In *Solenomya* the rectum is close to the ventral wall of the ventricle, instead of being connected with the dorsal wall, as is generally the case. (Pelseneer, No. 17.) In the forms in which the ventricle is not traversed by the rectum, *Nucula, Area,* and perhaps *Anomia* are primitive forms, while *Ostrea* and *Teredo* are among the most specialized of lamellibranchs. But in *Nucula, Area,* and *Anomia* the ventricle is dorsal to the rectum, while in *Meleagrina, Ostrea,* and *Teredo* it is ventral to it. In certain primitive mollusks (*Cephalopoda* and *Amphineura*), the ventricle of the heart is dorsal to the rectum, as in *Nucula,* etc., and it is probable that this is the primitive condition.

In *Nucula* and *Arca* the ventricle has the appearance of being double; that is, with distinct right and left halves. This has given rise to a discussion as to its meaning. Milne-Edwards (No. 12) regards it as the primitive condition, pointing to a double origin of the heart. Grobben (No. 6) considers the single heart as the primitive one and believes that this double condition has been brought about by the forward extension of the foot-retractor or byssus muscles. Theile (No. 22) believes that the heart was originally of two independent halves and that upon uniting in the median line the heart is found in its various positions, inclosing the rectum or lying above or below it. His conclusions are based on the fact that Ziegler (No. 24) has shown the pericardium of *Cyclas* to be formed in development from two symmetrical vesicles which unite in the median line. At about the same time that Theile's paper, above referred to, appeared, Pelseneer (No. 17) expressed the opinion that neither the views of Milne-Edwards nor those of Grobben were altogether correct. He discards the view of the latter entirely. He says that this double condition is not absolutely primitive, but is "due to the bilateral separation of the gills (much closer to one another in other lamellibranchs) and the auricles." The primitive heart, he believes, was dorsal to the rectum, and the commonly perforated heart was formed by a fusion of the two parts of its ventricle below the rectum. In *Ostrea*, when the heart is ventral to the rectum, this position was acquired by the great development of the single adductor; for, as it extended so greatly dorso-ventrally, it carried down the gills, and with them the auricles. These, not elongating, also compelled the ventricle to move downward. I had myself already come to this conclusion in regard to the peculiar position of the heart of *Ostrea*.

The pericardium is generally a large space in which the heart lies. Its relative size varies extremely, however, in different forms which may be closely related. For instance, the pericardial space of *Mya arenaria* (Fig. 27, *p*) is quite small, while in *Venus mercenaria* (Fig. 16, *p*) it reaches a very great size. Lacaze-Duthiers showed that the heart of *Anomia* did not lie in a pericardium, and he believed that none existed. Pelseneer, however, thinks that he has found the remains of the pericardial space in a flat cavity beneath the rectum, into which open the inner ends of the kidneys. The pericardium is not connected with the vascular system, but into it open the excretory organs, one on either side, and, when they are present, the pericardial glands.

In most cases the heart of a lamellibranch gives off an anterior aorta which runs forward above the rectum, and a posterior aorta proceeding below the rectum. In some forms, however, the posterior aorta does not appear. Such an exception is *Solenomya*, one of the most primitive forms.

It has been claimed that the possession of two aortæ was the primitive condition and that when the posterior one was absent it had degenerated and disappeared. As the general relations of such a form as *Solenomya* have become known they have shown that it is a primitive form; and, as it possesses but one aorta, that condition may have been the original one in the group. In the gasteropods there is no posterior aorta, but a branch of the anterior aorta supplies the posterior part of the body. The remaining groups of the Mollusca have both anterior and posterior aortæ. Pelseneer (No. 17) considers the single aorta the original condition in lamellibranchs.

As has been said, the oyster possesses both anterior and posterior aortæ. The former springs from the upper (morphologically anterior) end of the ventricle and runs forward to the visceral mass and mantle. The latter takes its origin slightly posteriorly to the first, very close to the extreme anterior end of the ventricle, and runs directly backward beneath the rectum and in the upper wall of the pericardial chamber. Upon reaching the adductor muscle it turns downward, running along its anterior surface to a little below its middle, and then penetrates its tissue and becomes distributed among the fibers.

The oyster probably came, in its degeneration, through a form with an anterior and posterior aorta. But in this case the posterior aorta must have shifted its position from the posterior end of the ventricle to its anterior dorsal end.

Arteries break up, not into capillaries, but into irregular blood spaces, in all tissues of the body which they penetrate. In the gill filaments, however, the blood channels are of more regular size, and in many cases (*Pecten*, Fig. 83) are easily seen to be lined with a distinct endothelium. In the walls of the digestive tract and in the labial palpi also, the blood spaces are quite regular, and much like definite vessels. These are lined by an endothelium, as is shown in Fig. 75, Pl. xc, *br*, a section of the mouth fringe of *Pecten irradians*.

The blood of lamellibranchs is colorless, with a few exceptions, and contains many corpuscles. Some of the *Arcas* and *Solen* (Lankester, No. 8) have corpuscles containing hæmoglobin, so that the blood is distinctly red.

The relative amount of the blood is very great in some locomotor forms where it is used in protruding the foot, and in such sedentary forms as *Ostrea* and *Mytilus* is comparatively small.

The course of the circulation is as follows: From the ventricle of the heart the anterior aorta conveys the blood forward along the dorsal wall of the visceral mass, over the stomach, and then down into the foot. From this main artery many branches are given off to the tissues of the liver, sexual glands, palps, digestive tract, and foot. Where the posterior aorta is present it is in most cases distributed mainly to the mantle folds, and also supplies the siphons of the mantle and the posterior adductor. If the posterior aorta be absent, these posterior tissues are supplied by a branch from the anterior aorta. From the irregular sinuses into which the arteries empty, the blood is collected in larger vessels and conveyed to a vessel beneath the pericardium, called the sinus venosus. Thence it passes to the gills, traversing on the way the walls of the nephridia, where waste products are excreted. The circulation is completed by the return of the blood from the gills to the auricles of the heart.

The path of the blood through the gill filaments is not well known and would be impossible to determine in those forms in which the gills have become greatly specialized, owing to their complex form. On account of the colorless condition of the blood corpuscles also, their movement in the filaments can not be followed.

It seems altogether probable that the manner of the circulation in the gills is very dissimilar in different groups of lamellibranchs. In those forms in which the gills are made up of a series of leaf-like plates (*Nucula, Yoldia, Solenomya*), each of these is little more than a blood sinus (*Yoldia*, Figs. 79, 80, 82; *Nucula*), around whose outer edge a blood channel is more distinctly marked out. While a circulation may be more or less distinct here, it can not be perfectly so.

In forms with a descending and ascending portion in the filament, and where the latter is not in concrescence at its extremity with the mantle or neighboring filaments, the blood must flow out to the extremity of a filament and then back again, perhaps, however, not taking exactly the same path. It will be noticed that in gills of this kind the blood space of the filament is divided by a septum or is greatly flattened out and shows in section a long, narrow blood channel. This latter condition is found in *Arca pexata* (Fig. 66, Pl. LXXXVIII).

The blood corpuscles of *Arca* are colored by hæmoglobin, and I have attempted to trace the circulation in the filaments. I removed one valve of the shell, placed the animal in water, and examined the gill by separating, without injury, a few filaments at a time, as they are connected with one another only by the ciliated junctions on their sides. Under the microscope the pale yellow corpuscles could be seen in motion in the blood stream. In both outer and inner lamellæ the currents would keep up a constant flow outward for half a minute, at times, though the rate of the current did not remain constant. The streams would become slower, finally stop, and then a back-flow would set in. After a short time these ceased, and the currents resumed their original course. I did not confound two separate currents, but could see individual corpuscles being carried in one direction and then back in the other.

The currents in the different filaments were independent of one another, and there seemed to be a somewhat irregular channel for them.

In cases where the filaments are joined to the mantle and the ventral side of the visceral mass (*Ostrea*) the blood may be finally collected in vessels running along the line of this concrescence. The currents through such filaments can not be at all definite, as the filaments open into one another at various places.

It seems probable that the gill is used as a respiratory organ in all lamellibranchs, though in a few very little blood, apparently, gets into them. In some cases the other surfaces of the body, and particularly those of the mantle, may play a more important part than the gills in the aëration of blood.

THE EXCRETORY ORGAN.

The nephridium is situated immediately beneath the pericardial chamber. The more primitive condition of the gland is preserved in *Nucula* and *Solenomya*, where it is a simple tube, bent upon itself. One end of this opens into the pericardium, the other to the exterior by the branchial chamber. The lining epithelium in these cases is similar in all parts of the tube and is all secretory (Pelseneer, No. 17). It consists of large cuboidal, vacuolated cells, without concretions. The generative gland opens into this kidney and near its pericardial opening. There are two excretory organs, one on each side, having no connection with one another.

There are many variations from this simple loop, but the general plan is always followed. In most cases the loop becomes differentiated, so that its terminal half becomes nonglandular, while only the half connected with the pericardium remains glandular. On account of the loop in the organ the glandular portion is ventral and the nonglandular is dorsal in vertical sections. The gland is seen in section beneath the pericardium in Fig. 16, Pl. LXXXII. *gl* represents the large glandular portion and (*ngl*) the nonglandular tube above it.

The glandular portion possesses greatly folded walls, bounding its lumen, for the increase of the secreting surface. In *Anodon*, Rankin (No. 20) has described the lining cells as being more or less ciliated.

In *Pecten*, the nephridia are situated beneath the adductor muscle (Fig. 46, Pl. LXXXVI, *n*) and are exposed to the exterior, as they hang in the mantle chamber. They are connected with the ventral floor of the pericardium on either side anteriorly, and this opening on one side is shown in Fig. 45, *n*, the adductor muscle here having been entirely cut away.

The position of the organ in *Ostrea* is seen in Fig. 97, Pl. XCIV, at *n*. Its anterior end lies under the pericardium, and it extends backward, close under the adductor, as far as the small notch between the light and dark muscle fibers. It is thus seen, as in *Pecten*, to expose a great part of its wall directly to the water in the epibranchial chamber. The glands of either side are entirely separate from one another. The secreting portion of the gland gives off a great number of branching diverticula, lined by an excretory epithelium.

A portion of one of these diverticula is represented in Fig. 54, Pl. LXXXVII. The lining cells (*exe*) are columnar and rest upon a thick basement membrane. The cell protoplasm is mostly collected at the bases of the cells. The distal ends contain large, watery vacuoles which are finally discharged into the lumen of the tube (*l*). Many of these vacuoles contain a small, round particle which stains deeply and which may be of a concretionary nature. Concretions as commonly found in other forms, however, seem to be entirely wanting here. The products discharged into the lumen are undoubtedly liquid, for by tapping on the cover, when fresh preparations are used, they may easily be made to run into one another and form larger masses. None of these cells give evidence of ciliation.

Figs. 56 and 57 represent concretions from the secretory cells of the kidney of *Pecten*. They are sometimes very large and their concentric structure is easily seen. Figs. 58 and 59 are from macerated preparations. The former shows a basal nucleus (*n*) and the concretion in the distal portion of the cell (*con*). The concretions possess a more dense, deeply staining, central portion of varying shapes and sizes. Scattered through the cell substance are many small, vacuole-like bodies (*vac*), probably the first appearance of the excreted products in fluid form, which afterward construct the solid concretion. Other cells (Fig. 59), of a more elongated shape, appear to contain only watery vacuoles (*vac*).

The secretory cells are deeply pigmented in most lamellibranchs, but are entirely devoid of coloring matter in the oyster. The gland is thus rendered very inconspicuous and was for a long time overlooked. It has been described in the European oyster (*Ostrea edulis*) by Hoek (Journ. Soc. Néerlandaise de Zöol., 1883).

By means of this tube-like nephridium, the pericardial cavity is connected with the exterior, and it has often been supposed that water entered the pericardium by this means. This has, however, never been demonstrated, and there are reasons for supposing that it does not occur. Among these are the facts that the narrow channels leading from the pericardium and to the exterior possess, as a rule, many cells with very great cilia, which have been shown, in some cases, to cause a current outward; and also in *Pecten* there are valves guarding the opening into the pericardium, preventing any such back flow into it.

Garner has shown that the sexual gland of *Pecten* opened into the kidney. In examining the nephridia of this form in the spring and early summer, I have almost always found that they contained eggs in various stages of development, often in very great numbers. They are, in all probability, the eggs of the individual examined, which have lodged in the kidney instead of being thrown out to the exterior, and not eggs from some other mollusk taken with the surrounding water. In this case they have probably become fertilized from spermatozoa taken into the kidney from the exterior, having come from another individual, if it is always true that the male elements in these hermaphrodite forms are always ripe before the female, as observations seem to indicate. The great majority of eggs are of course discharged into the water, and there become fertilized and develop.

It thus happens that the spacious kidney of *Pecten* becomes a brood pouch for some of the young, and I have seen a kidney containing great numbers of eggs far advanced in segmentation, which, it seems probable, both on account of their number and the connection between kidney and sexual glands, developed there from the beginning. It is true that this arrangement may be a purely accidental one, but I am inclined to believe that the advantage gained in protecting so many embryos for so long a period in their development is considerable, and that on this account the kidney may, by the operation of natural selection, have become especially adapted for this function.

THE NERVOUS SYSTEM.

Of the three pairs of ganglia the cerebral are usually situated near either side of the mouth and are joined by a supraœsophageal commissure. They are generally close to one another in the more primitive forms. The visceral ganglia, generally placed in the ventral side of the posterior adductor, occupy in *Solenomya*, *Nucula*, and a few others, a position in front of it. Connected with the pedal ganglia is the otocyst. Though it often lies on the surface of the latter, it is probably always innervated from the cerebral ganglia. In some cases the otocyst contains a single otolith, in others there may be several small particles.

Sections of the otocyst of *Yoldia* show an epithelium, the cell walls of which could not be distinguished. Nuclei of different shapes and sizes are scattered about irregularly through it. These cells rest upon a dense supporting membrane, and the whole is enveloped in a capsule possessing a fibrous appearance. There is but a single large otolith, whose concentric structure is very evident. There is also in the cavity of the otocyst what appears to be a coagulated fluid. The lining epithelium, as far as I was able to see, possessed no trace of cilia. This apparent absence of cilia has also been described for the otocyst of *Nucula*, by Pelseneer (No. 17).

THE GILLS.

The gills, of which there are four, hang in the branchial chamber. They are represented in Fig. 96, Pl. XCIV (*Venus mercenaria*), the right valve and mantle lobe having been dissected off. The line of attachment of this mantle lobe dorsally, beginning at the posterior end at the side of the siphons, is upward along the back of the posterior adductor; thence it proceeds forward in a curved line near the top of the pericardium and visceral mass to the anterior foot-retractor, which pierces it, and over the anterior adductor.

The free edge of the left mantle lobe is seen at the point *me*. An extremely large branchial chamber is thus formed, the upper boundary of which is the line of attachment between the mantle and outer gill. In it, over the walls of the visceral mass, hang the gill plates (*og* and *ig*), and below is suspended the foot. The gills of *Ostrea virginiana* (Fig. 97) extend from near the anterior end of the body backward, and up for some distance on the posterior side. In *Pecten* their extent is so great as almost to surround the entire adductor muscle and visceral mass. In *Ostrea*, each of the four gills is connected above with the ones next to it, and the outer ones are also connected with the mantle. A vertical section (Fig. 2, Pl. LXXIX) shows each gill to be made of two lamellæ (*ol* and *il*), leaving a space (*w*) between them. They are united to each other above, the top of the inner lamella of one to the top of the outer lamella of the next, and on the median line the opposed inner lamellæ. In Fig. 2, which is a section near the anterior end of the gills, they are also all united above to the visceral mass. A little farther back, however (Fig. 4), the two inner lamellæ of the inner gills only are attached to the body above. Farther back still, under the adductor (Fig. 8), the gills are entirely free from the parts above. In all regions, however, the outer lamellæ of the outer gills are united with the mantle, as may be seen in all the figures of cross-sections of this form.

By this concrescence of the gills above, the mantle chamber below is completely shut off from all the spaces which appear in section above them. Posteriorly these epibranchial chambers open into one another in one large cavity (Fig. 8, *c*), forming a cloacal chamber. Its position on the posterior side of the adductor muscle is shown at *c* (Fig. 97, Pl. XCIV).

The gill lamellæ are made of innumerable parallel filaments united to each other in various ways in different forms, and always leaving openings through which water may enter the epibranchial from the branchial chamber. This current is caused by the ciliated cells of the filaments. The spaces between the lamellæ are called water tubes. The currents from the epibranchial chambers pass posteriorly to the cloaca. Into it, from above, opens the rectum in all cases (Fig. 97, *r*). In those forms which possess siphons, the cloaca opens into the anal siphon (Figs. 93 and 94).

The more common position of the gills differs from that of the oyster, in that a foot is generally present, situated on the ventral side of the visceral mass and protruding between the inner gills. If this should occur in *Ostrea* and the gills should then be moved up on the sides of the body, we would have the condition in *Venus* represented in Fig. 14. The whole foot and visceral mass here separate the chambers of the right and left sides, instead of their being side by side, as in the former case. Behind the visceral mass and foot, the inner gills join one another, forming a branchial septum, which is continued posteriorly to the base of the siphons, and still preserves a complete separation between the cloaca and branchial chamber (Fig. 93, Pl. XCIII).

Here also, as in *Mya* and many other forms, the epibranchial chamber is divided anteriorly into four parts, two on each side of the body, farther back into two, and finally these unite into one, the cloaca.

In *Mytilus* there are no sharply defined epibranchial chambers, for, as may be seen in Figs. 32 to 41, the inner lamellæ of the gills do not fasten to the body wall, nor do the outer lamellæ unite with the mantle as they do in the forms just noticed. It thus happens that the water tubes (*w*) open directly into the branchial chamber, whence their supply is obtained. The backward current is, however, confined to the dorsal part of the branchial chamber and leaves it by a special siphonal opening at its posterior extremity. (Fig. 87, Pl. XCIII, *co*, a view of the mantle edge in the posterior region.)

Filamentous gills of this sort, often, however, undergoing great complications, are possessed by the great majority of lamellibranchs. The gills of a few primitive forms (*Nucula, Yoldia, Solenomya*), however, are entirely different in appearance. They were first made known by Mitsukuri (No. 13). Instead of filaments, each gill is made up of a number of flat plates, placed one against another, the two gills on each side being supported by a common muscular membrane, which is attached in the usual position to the sides of the visceral mass. The palps in *Nucula* and *Yoldia* are very large and extend back for some distance upon the visceral mass. The gills, posterior to them, extend backward to the posterior end of the body. Fig. 92 represents half of two gills of *Yoldia*. The cut surface at the right of the figure exposes not one continuous plate, but two plates, one on each side of the supporting membrane *m*. On the ventral side of the gill, opposite the supporting membrane, is a groove, not well shown in the figure (*gr*), separating the plates of either side. Other evidence will be given to show that these plates are really separate, and one plate does not extend entirely across the gill.

This organ in *Solenomya* is very similar to the one described. Here (Fig. 91) the gill on either side is attached to the visceral mass by a short supporting membrane in such a way as to almost completely envelop the posterior part of the body. The outer plates of the gill now extend upward on the side of the body instead of hanging down below the point of the attachment of the supporting membrane, as in *Yoldia*. The plates of the upper row differ in shape from those of the lower, being longer and narrower.

There is yet another condition of the respiratory organs of lamellibranchs, first described by Prof. Dall, of the Smithsonian Institution (No. 4). In *Cuspidaria* and *Poromya*, probably very far removed from primitive forms, the gill as such seems to have disappeared. On either side of the body, extending from the walls of the visceral mass out to the mouth, is a thick, horizontal, muscular membrane. It extends the entire length of the animal, from close behind the anterior adductor back to the siphonal septum, and separates the mantle chamber on either side of the body into an upper and a lower chamber. Through this membrane open a number of orifices of various arrangement, which allow a passage of water upward from the lower into the upper chamber. This latter corresponds to the usual epibranchial chamber, water obtained from the chamber below being discharged through the anal siphon. The upper chamber is stated by Dall (No. 5) to be used as a marsupium.

There seems to be a question as to whether or not the gill has disappeared in those forms and whether the muscular membrane is homologous with the gills or a morphologically different organ. Dall (No. 4) some years ago expressed the view that this membrane was not morphologically a gill, but that it was a great extension anteriorly of the muscular siphonal septum found greatly developed in other forms. A gradual transition, in which the true gills become smaller while the siphonal septum increases in area and extent, is traced through the forms *Lyonsia*, *Lyonsiella*, and *Verticordia*, all these forms possessing true gills. With the loss of the gills the septum takes upon itself their function of respiration, and the progress of its specialization for this purpose after the gills have disappeared is illustrated in the series of forms *Myonera*, *Cuspidaria*, *Ctenoconcha*, and *Poromya*. There are possibly cases in which the muscular septum is made up of structures diverse in their origin, the anterior part being from the gills and the posterior from the siphonal septum.

More recently Pelseneer homologized this septum with the gills (No. 17). The reasons given for this view are that, while it is connected with the siphonal septum, the latter has a different innervation, while the muscular septum proper is innervated by the branchial nerve, and that it is in direct communication with the efferent lacunæ of the auricles.

DETAILED STRUCTURE OF THE GILLS.

The gills, so extremely varied in structure, are of very great importance on account of the relationship which they show to exist among various forms of the group, and have been thoroughly studied in a great many genera. The work of Posner (No. 19), Peck (No. 16), and Mitsukuri (No. 13), extended over a great field, and, together with the embryological observations of Lovén (No. 11) and Leydig (No. 10), has established a fairly satisfactory view of the phylogenetic history of the gill. Mitsukuri, who first made known the structure and nature of the gills of *Nucula* and *Yoldia*, now considered to be the most primitive of living lamellibranchs, was the last of these authors mentioned to publish his views; and he reviewed the work then completed, giving a theory of the phylogenetic development of the gills, which is generally accepted as the true one. He says: "To review the whole matter, the lamellibranch gill was perhaps originally a simple ridge on the side of the body, but, to increase the surface of contact with the water, folds may have arisen on two sides of this ridge. If such was the case, *Nucula* and *Yoldia* are still in a stage very little advanced from this primitive condition. In course of time, however, as some of the Lamellibranchiata, either owing to degeneration or some other cause, became incapable of extensive locomotion, these gills or folds were perhaps prolonged to form tentacular filaments, which, going on in their development, finally produced such complex gill structure as we see in *Mytilus, Unio, Ostrea,* and other forms, taking on at the same time functions totally foreign to their original one."

THE GILL OF YOLDIA.

The primitive plate gills of these two forms, already briefly described, were first studied by Mitsukuri, whose attention was directed chiefly to the former. On account of poor material, he was not able to examine into the histology of the gill of *Yoldia* in any detail, though he gave an account of the more important features of its structure.

As described by Mitsukuri, the plates of the gill are suspended from above by a thick membrane (Fig. 92). Close to its attachment to the gill plates, there is a blood channel running the length of the gill, and a similar vessel is present also in the median line, just above the groove on the ventral side of the gill, which separates the plates of either side from below. The course of the blood in these channels is not known with certainty. Mitsukuri made very little out of the histology of the plates.

Contractions.—If a living *Yoldia* be removed from the shell and examined in sea water, the gills will be seen to possess a deep-red color. The thickened ventral edges of the plates seen from below are light red, but the thinner lateral edges are of a much darker hue. The gills will be observed to be able to contract themselves in a variety of ways and to be very susceptible to stimulation from without. In the first place, they may shorten themselves to a considerable extent from before backward, and, like all the movements of the form, this may be done very quickly. A contraction may also take place in such a way as to greatly reduce the circumference from side to side and from above downward. This last contraction is a very common one and at times occurs in a curious way. At any point in the gill, three or four plates adjoining one another, and the same number on the other side, opposite these, may contract themselves apparently from every direction, thus giving the appearance of a deep groove running entirely around the gill. Now this contracted zone begins to move along the length of the gill and it may move in either direction. In this wave of contraction but three or four plates on either side are ever concerned at one time. These waves are often single, and at times several may follow one another in succession.

Still another wave contraction may often be noticed on the ventral sides of the plates. The ventral side of a single plate, or at most two plates, is affected at one time. These waves occur independently on either side of the gill. A single plate bends a certain region of its ventral surface forward or backward, so as to separate this region from one of the neighboring plates and bring it close to the plate on the other side, either before or behind it, as the case may be. This latter plate quickly bends in the same way, the first one assuming its original position and then the succeeding plates, thus causing a wave of this bending to run along the length of the gill. These waves may run either forward or backward. Single plates may contract slightly independently.

If a gill be dissected out from the body, these contractions still continue to take place whenever it is touched. The action of the cilia is so powerful that the entire gill is made to move about in the water.

Collection of food.—Mitsukuri came to the conclusion that this primitive form of gill was concerned only in the aëration of the blood, and that it was probably not concerned in the procuring of food. I was able to observe in the case of *Yoldia*, however, that not only was the function of gathering food possessed by the gills, but that it was performed with amazing rapidity. Carmine particles in the water once coming in contact with the ventral edges of the plates, having been swept there by the powerful currents which these ciliated borders set up, are at once hurried along toward the wide, median, ventral groove of the gill, into which they are thrown. On the way to this groove they have evidently become covered by mucus secreted by gland cells; for the separate particles of carmine are soon firmly cemented together, and passing along the groove anteriorly, less rapidly than on the edges of the plates, though still at a comparatively fast rate, they are finally piled up at its anterior end and gradually passed upon the surface of the palp.

Structure.—A diagrammatic view of the ventral side of the gill is given in Fig. 78, Pl. XCI, in which the thickened ventral edges of the plates (*p*) are shown on either side of the groove (*gr*). These edges in the living gill are only slightly bent downward, and are not so curved in outline as represented in Fig. 81, which was drawn from a hardened specimen. From the groove (*gr*, Fig. 78) to the point *cj*, the plates are in no way connected with one another, and one may see entirely through the gill between them, when examining from below.

At the point *cj*, where the edge of the plate turns abruptly upward, occurs a ciliated junction not before described in these forms, between adjoining plates. The cilia from neighboring plates interlock closely, making a comparatively substantial union. This ciliated junction is not confined to the point mentioned, but extends from it upward on the lateral sides of the gill for about two-thirds of the distance to the upper pointed extremities of the plates, the ciliated union being confined to the lateral edges of the plates.

An examination of sections made through the gill plates in different regions will give a better idea of these points. Fig. 79 represents a section cutting the plate of the gill of *Yoldia* in a plane represented by the line *a* in Fig. 81. This section thus shows the structure of both the ventral and the dorsal borders of the plate as well as that of its interior regions. The thick ventral border is shown at the upper end of the figure. As pointed out by Mitsukuri for *Nucula*, this corresponds in structure to that which is usual in the outer edges of the gill filaments of other lamellibranchs. This portion of the gill is supported by a framework of the so-called chitin (*ch*), which widens out at about the middle of its extent and makes the inclosed blood space a conspicuous channel at this place (*b*). The "chitin" is here quite thick and gradually becomes thinner in either direction.

The outer epithelium of the ventral edge is made up of closely packed columnar cells. Those at the extreme edge of the plate (*e*) bear peculiarly long and powerful cilia, which differ greatly in appearance from cilia in other regions of the ventral edge of the plate, or in fact in any other gill. In cross-section they are circular. The pencil cells, found frequently in other regions of the body in lamellibranchs, bear a stiff bristle which is found, by maceration, to be made up of several fused cilia. Whether or not these cilia may be of a similar structure I can not say, but it seems hardly necessary to suspect such a structure merely on account of their great size. Judging from their position and that of the other lines of cilia on the edges of the plate, these greatly developed cilia of the frontal cells are the ones which produce the rapid currents in the water over the ventral surface of the gill.

At some little distance inward from these frontal cells, the epithelium rises into a ridge on either side (*r*) and these bear a second row of cilia much shorter than the first, and very fine. They protrude laterally and outward, and their ends touch those of similar rows of cilia on contiguous plates. These rows, however, do not interlock with one another, and I believe that they serve simply to prevent currents of water, bearing food particles and other foreign bodies, from getting in between the plates, and not at all as a means of connecting neighboring plates. I think, also, that similar lines of cilia on the filaments of other forms, which will be noticed, serve the same purpose.

A third line of cilia is borne by elongated cells, which do not, however, form a ridge, and is situated near the inner edge of the thickened, ventral edge of the plate (*t*). These cilia are longer than those of the second row and are also fine in appearance. In sections they appear bent outward toward the edge of the plate.

All these rows of ciliated cells are sharply defined and the cells between them bear no cilia. Gland cells usually exist in the frontal region of all lamellibranch gill filaments or plates. If they appear in this region in the gill of *Yoldia*, it is in very small numbers. I have frequently seen cells here which appeared much like gland cells, but I have not been able to decide positively that they were such.

Lining the blood space inclosed by the chitinous layers is a perfectly distinct endothelium, represented in the figure. Not only were the slightly elongated nuclei easily seen, but also the cell protoplasm flattened out over the chitin. There was no possibility of confounding these cells with the nucleated blood corpuscles seen in the blood space, as certain observers have been accused of doing in other forms.

The thickened ventral edge, as seen in the section, is sharply separated from the remainder of the plate. This latter portion is made up of large epithelial cells, whose boundaries are not distinct and are seen but occasionally. These cells are of uniform size throughout, and through them is evenly distributed a light yellow pigment, which is quite abundant and probably gives the red tint to the living gill. Immediately above the ventral edge, these sides expand laterally, leaving a large space (s), in which are numerous blood corpuscles. At the dorsal edge of the plate (d) there is also a marked widening of the bounding walls. The walls in the center of the plate are found in sections to be greatly convoluted (f), probably due to the contraction of muscular fibers (mf) contained within the space between the walls, to be described later.

In these three regions described, the walls are connected with one another by numerous, branching, lacunar cells (lac), though they are less numerous in the central region (f).

The entire interior of the plate is seen to be of one continuous blood space, with definitely constricted areas and enlarged channels. While blood corpuscles are naturally more abundant in these channels, they are found in all parts of the interior of the gill plate.

Fig. 82 represents a section cut horizontally through the gill plates of one side in a plane indicated by the line b in Fig. 81. In this region, above the thickened ventral border, no chitinous framework appears. The outer edge of each plate is rounded and composed of short columnar cells, bearing very short, fine cilia on the extreme outer or frontal surface (e); but on the lateral sides of the rounded edges are narrow lines of cilia interlocking closely with each other (cj). At times, spaces (s) are found where the ciliation is absent, though these may perhaps be due to a mechanical tearing of one plate from another. The extent of these ciliated lines on the sides of the plates has been referred to above. The remaining parts of the plate walls are the same as already described, though less convoluted. In this section a part of the supporting membrane of the gill has been cut across. Its surface epithelium (ep) is columnar, and many of the cells bear thick spines, perhaps bundles of cilia. The inner part of the membrane is made up chiefly of muscle bundles (m).

Fig. 80 represents another horizontal section, passing close to the dorsal end of the plate (in the plane of c, Fig 81). Here the ciliated connecting lines or rows have disappeared and the epithelium is everywhere the same pigmented, indistinct kind above described for the main body of the plate. In this dorsal region, the lacunar cells (lac) are very numerous, and their processes, extending from one wall to another, are generally fine and thread-like.

That the plates of one side of the gill structure are not continuous with those of the opposite side is evident from a section passing horizontally in the plane indicated by d in Fig. 81. Such a section is represented by Fig. 48, Pl. LXXXVI. When the plates come together from either side on the median line (ml) they are not opposite each other, but on the contrary, break spaces. They are figured by Mitsukuri as being opposite each other in Nucula. The space a represents the interior of a plate, and b the space between two plates. The wall of one plate, then, runs over on the median line of the gills and becomes continuous with the wall of the next on the same side.

Fibers have been described in this gill by Mitsukuri as running down into the plates from the supporting membrane above. He regarded them as chitinous structures, serving to keep the plate expanded for purposes of aëration. They are shown in Fig. 79 and Fig. 48, mf, as being cut across more or less transversely. They are always closely applied to the inner face of one of the walls of the plate, but whether always to the anterior or posterior surface I do not know. A longitudinal, nearly vertical, section, which passes through the supporting membrane, shows these fibers to be continuous with its musculature. I believe them to be muscles and the means employed to bring about the numerous contractions described above as taking place in the living gill. They do not penetrate the thickened ventral portion of the gill plates, nor do they seem to be present in their extreme dorsal portions; their insertion seems to be upon one wall of the plate and at all points throughout their length.

THE GILLS OF SOLENOMYA.

The external appearances of this gill have already been mentioned. The morphologically ventral surfaces (here lateral in position) are slightly enlarged, but not nearly so much so as in Yoldia. These specialized edges extend from the median ventral groove of the gill to the pointed outer ends of the plates.

The ventral edge of the plate is composed of entirely similar cells throughout, resting upon the usual chitinous layer (Fig. 77, *re*). In the preparation from which the section represented was made, these cells had shrunk somewhat and separated from one another so as plainly to show their structure. They are columnar, have a finely granular protoplasm, the outer edge of which is deeply stained, and each cell bears several cilia. The cilia of all the cells are of the same length, excepting those on the extreme ventral border, where they are slightly shorter. At the lateral edges of the ventral border—never in the midline of the ventral surface—open certain gland cells (*gl c*), which are constantly and generally easily seen in a corresponding position throughout the gills of lamellibranchs.

Pelseneer (No. 17) figures two cells in the section of each plate (in the region just below the gland cells in Fig. 77), which are larger than the others and bear much longer cilia. These he probably means to compare with the "latero-frontal" cells of Peck, not so widely found among lamellibranchs. I believe, as seems to be so generally supposed. I am confident that these large cells do not exist in *Solenomya velum*. Pelseneer possibly mistook a gland cell in this region for a large nucleus and supposed it to indicate a large cell.

It is possible, though I think not probable, that the cilia on the sides of this ventral edge are used for joining the plates to one another, as described by Pelseneer. If such a junction exists it is a very slight one, for the plates, unlike those of *Yoldia*, are very easily separated. No direct interlocking of cilia is anywhere seen in sections, as figured for *Yoldia* in Fig. 82. The ends of the cilia of neighboring plates merely touch each other (Fig. 77), and, as in other cases, I believe that they function principally in keeping food particles from entering between the plates, thus confining them to the ventral surface, where they may be rapidly swept to the median ventral groove of the gill and onward toward the mouth.

The chitin of the gill plate is thickest at some distance from the extreme ventral edge (*ch*), and it is here relatively thicker than in *Yoldia*. It extends, in a long, thin layer, entirely to the ventral edge, and also in the opposite direction, toward the center of the plate. This latter extension rapidly thins out and disappears. No very marked expansion of the blood space takes place between the chitin plates.

The epithelium of the walls of the plates is much like that of *Yoldia*, though the cells are distinctly marked off from each other (*w*). They are more elongated and their outer ends are rounded. These cells also contain a fine, equally distributed pigment in their protoplasm. Sections show no indication of the folding of the walls, such as is found in the gill of *Yoldia*, and I have seen no trace of the muscle fibers which I have described as being present in the interior of the plate in the latter form. Endothelial cells are easily demonstrated, lining the interior of the chitinous layers (*ch*). Lacunar tissue, if it exists, was not made out, as the walls of the plate are everywhere closely applied to each other, in sections, and only here and there can very narrow spaces be seen between them.

THE GILLS OF ARCA PEXATA.

If an entire filament of the gill of this form be isolated, it will be seen to be made of a fully developed descending and ascending limb (Fig. 50, Pl. LXXXVI). The upper end of the ascending limb in the filaments of both gills is free from the mantle or side of the visceral mass, as in *Mytilus*. It bears an enlarged triangular end plate which turns outward. The anterior and posterior faces of this end are ciliated and form a large patch which closely interlocks with those in the same position on neighboring filaments.

The most striking fact in regard to these filaments is that the descending and ascending limbs are not separated from each other for about half their extent—the ventral half—but are connected by the continuation of their inner walls (*cp*, Fig. 50, Pl. LXXXVI, and Fig. 66, Pl. LXXXVIII).

There are no vascular connections between the filaments, but they are held together, as is usual in the genus *Arca*, by ciliated disks (*cd*. Fig. 66) arranged in a row on the anterior and posterior faces of the filament throughout its entire extent.

Fig. 66 represents a cross-section of four filaments. Three of these are cut at a point above the union of the ascending and descending limbs: a single one is cut lower down and passes through this connecting portion (*cp*).

The sections show the filament to be thin from before backward and wide from side to side. The elongated cells of the frontal epithelium (*f*) are uniformly ciliated and extend back for some distance from the end, the cilia gradually becoming shorter, as do the cells which bear them. At a corresponding point on either side, there are three or four cells seen in each section, which are longer than usual and bear long cilia (*lfl*). These are long ridges, or rows of cells, which are here cut in transverse section. They extend the whole length of the filament without any break, and their cilia do not serve to join contiguous filaments, but, as I believe, like the entirely similar rows in *Yoldia* (where there are two), are merely to prevent the main currents, carrying foreign particles, from entering between the filaments. Of course, currents of water do penetrate between the filaments to enter the water tubes of the gills, and thence proceed upward into the epibranchial chambers; but in all cases, so far as I have observed, the cilia of the rows, or lines, point obliquely *outward*, presenting their ends to foreign particles and keeping them out on the ciliated frontal epithelium while not being thick and heavy enough to prevent water from entering.

Peck (No. 16) has shown that in a position similar to this, though nearer the outer edge, in the gill filaments of *Anodon*, there was a *single* line of enlarged cells bearing very long cilia. Moreover, in *Dreissena* he found *two* lines of enlarged cells lying together, with no cells between, but with the two lines of cilia distinct. These cells, one appearing on either side in each section, he called the latero-frontal cells. In the latter case he distinguished a first and second latero-frontal cell.

The large ciliated cells, then, occur either in single rows or, I believe, in the majority of cases, in compound rows, showing several closely packed cells in cross-section. They do not always occupy a latero-frontal position. Considering these reasons, and the supposed function of their cells, I would designate them as the simple and compound straining lines.

In the case of *Dreissena* there are two simple, in *Yoldia* two compound, lines.

The inner part of the filament is composed of low cells whose boundaries are indistinct, excepting where an interfilamentar junction appears (*cj*). These are round patches of columnar cells bearing long cilia, arranged in the same region of the filament, at intervals throughout its length. The patches appear opposite each other on neighboring filaments, their cilia closely interlocked, and form a comparatively firm attachment. The cilia of these connecting discs are too closely packed together to allow of much motion, but by a high magnification, whether locked together or torn apart, the cilia of these patches show in a very feeble way the characteristic lashing in one direction and then the slow recovery of position. This motion is very slow indeed, each cilium moving independently of the others. This motion is of service, probably, in reuniting patches which have been separated from one another.

The chitinous lining (*ch*) is thin and at no place shows a marked thickening. It is slightly thicker just interior to the straining-line cells (at *ch*).

The membrane connecting the two limbs of the filament possesses walls, each of which is, as usual, made of a single layer of cells. They do not differ much in appearance from those of the inner part of the filament proper, with which they are continuous, except that they are slightly longer. These cells possess no cilia.

There is probably a continuous, vascular space between these walls, though in section they are, as in *Solenomya*, closely applied to each other. Here and there (*vs*) this space may be recognized.

THE GILL OF PECTEN IRRADIANS.

The gills of *Arca*, just described, and those of many other lamellibranchs, possess a smooth outer surface. In many forms, however, the lamellæ, both outer and inner, are thrown into definite folds, in order that their surface may be increased to facilitate aëration of the blood or the procuring of food, or both.

This folding appears in the lamellæ of the gills of *Pecten* (Fig. 86, Pl. XCII). The filaments are joined to each other only by cilia. The filament which occupies the salient angle of the fold (*sa*) is similar to the others in this form, although in some other cases (*Lima*, etc.) it is much enlarged. The filament at the reëntering angle in *Pecten* (*ra*) is greatly enlarged and peculiarly modified.

Fig. 86 represents a section passing through an entire fold of the gill. This fold is made up of fifteen filaments, not counting the specialized ones at the reëntering angle. The number varies slightly in different folds. All of the descending limbs of the filaments, forming one lamella of the gill, are cut on one side, and all the ascending, forming the other lamella, appear on the opposite side.

A section of a single filament is shown in Fig. 83. The chitinous layer is of nearly equal thickness, being thinner at the outer and inner edges, and thickest beneath the straining-line cells. The exterior epithelium is of long, large cells in the outer half of the filament, gradually merging into short cells, which are considerably widened. At the inner edge of the filament they again become columnar, but are not so long as at the outer edge.

Gland cells (*glc*) are present at the lateral boundaries of the frontal region (*f*) and are very conspicuous in certain specimens. They always occupy this position, and I have never seen them in the middle of the frontal epithelium, and only occasionally in other parts of the filament. These mucous cells vary in shape in different forms, sometimes being spherical, sometimes long and slender. They may be close to the base of the epithelium, surrounding them or pouring their contents out upon its surface. These various shapes and positions, more or less common in all lamellibranch gills, are shown in Fig. 85, a representation of the *inner* edges of four filaments of the gill of *Anomia*.

The cells of the epithelium immediately surrounding the gland cells in *Pecten* are seen to be crowded with yellow pigment granules (*pg*, Fig. 85).

At some little distance inward from the gland cells, appear in section four or five cells, bearing long cilia, crowding together to form the single compound straining line (*lfl*) on either side. These lines of cilia are inclined outward.

Fine cilia exist upon the frontal surface (*f*), becoming gradually shorter and finally disappearing as they reach the region of the gland cells. It is the movement of these cilia which causes the currents of water over the surface of the gills.

The chitinous layers, between which is the blood space of the filament, are lined on their inner surfaces by a distinct endothelium. Both the nuclei and protoplasm of its cells may be plainly seen (*en*). The blood channel, in which appear numerous nucleated blood corpuscles (*bc*), is divided by a transverse partition. Pelseneer describes this septum in *Pecten opercularis* as being chitinous in nature (though he does not use that term). I believe this septum (*p*) to be made of cells continuous with the endothelium of the walls of the blood cavity. Nuclei are very frequently present in it, often larger than the one represented in Fig. 85.

In the form referred to, Pelseneer also describes two large cells, one on either side of the frontal region—the latero-frontal cells of Peck—with their long cilia. Nothing of the kind exists in *Pecten irradians*.

The filament of the reëntering angle (Fig. 86, *ra*) is much enlarged. Its filamentous nature is recognized by the chitinous layers (*ch*), the compound straining line of ciliated cells (*lfl*), and the blood channel (*bc*). The epithelium of its walls is of a nearly uniform appearance. The descending and ascending limbs are connected throughout their entire length by a continuation of their walls (*w*), which thus separate the interior space of the gill between the lamellae into a single channel for every fold. The walls of the partition are generally applied to one another in sections, but very frequently spaces are found between them which contain blood corpuscles. I believe that a vascular connection exists here between the two gill lamellae, though it is said by Pelseneer not to exist in *P. opercularis*. Lacunar tissue also seems to be present (*lac*).

The ciliated junctions are peculiar and, so far as I know, are found only in *Pecten*. *Mytilus*, *Area* (Fig. 66), etc., possess filaments joined to each other by ciliated discs, situated on the sides of the filaments. In *Pecten*, however, the ciliated junctions do not occur upon the sides of filaments, but only upon conical projections from their inner surfaces. These are shown in Fig. 84, a diagrammatic representation of three filaments (*fil*) in a region where the projections occur. The conical points, which I would designate as ciliated spurs (*cs*), protrude abruptly into the water tube between the lamellae of the gill.

A section passing transversely through the filaments, and in the long axis of a few of these ciliated spurs, is shown in Fig. 86, at *cs*. Their walls are seen to be merely a continuation of the single-cell-layered wall of the filament. They are closely interlocked by cilia. Their interior is filled up with a solid mass of cells whose nature I have not positively determined. In macerated specimens, they appear to be much elongated in the long axis of the spur, but sections failed to show this satisfactorily. These projecting spurs are very conspicuous in any macerated preparation of the gill of *Pecten*.

GILL OF VENUS MERCENARIA.

This folded gill is, in many ways, more complicated than that of *Pecten*. The filaments have a vascular connection with one another at their inner edges, spaces being left here and there to allow water to enter the water tube of the gill (Fig. 70, *w*, Pl. LXXXVIII). The great primary folds, marked off by the primary reëntering angles (ra_1), include about seventy filaments. These folds are sometimes divided, however, by a second reëntering angle (ra_2), into two secondary folds. There is no connection between the gill lamellae at this point. At the primary reëntering angle there is a partition between lamellae, consisting principally of muscle bundles (*mus*). Between these exists a blood channel (*br*). From these channels is sometimes given off into the water tube on either side of the partition a huge blood sinus (*pbs*). These are so large as to almost completely fill the space between the lamellae.

Similar blood sinuses appear in other cases from an enlargement of the filament at the angle *fbs*, and may also be so great as to entirely fill the water tube. In regions where this condition is present, blood sinuses from the partitions are very small or absent.

Venus being a large form and quite active, must require a considerable amount of food and at the same time sufficient aëration of blood. The narrow filaments, not sufficiently large to contain much blood, seem to have been specialized for procuring food. These large blood sinuses may have been developed to provide for the diminished aëration in the filaments. Though they nearly fill the water tube, their thin walls are surrounded by water on all sides, probably in quantities sufficient for the purpose.

THE GILL OF OSTREA VIRGINIANA.

The gill of this very degenerate form is probably the most complex in the group. The lamellae are thrown into a number of folds between each of the thick cross-partitions. The filaments may be seen almost everywhere in section (Fig. 76, Pl. XC) to have a vascular connection with one another at their inner edges. This is the case with all the filaments in the fold marked 2 in the figure to which reference has been made. Here the entire inner space of the fold is a blood sinus, with which the blood channel of every filament is continuous. In fold 1, three filaments at its outer extremity are connected by their inner edges. In fold 3, four filaments on the left are thus connected, and this common sinus is seen to be continuous with another, close under the wall of the water tube of the gill (*wt*).

Very frequently, however, openings appear between the filaments. In fold 3 this is the case with the majority of the filaments in the plane of this section. Here water may enter between the filaments into the interior of the fold and thence into the water tube of the gill (*wt*).

A filament has much the same structure as in *Pecten*, excepting that the chitinous layer is abruptly thickened a short distance from its outer edge (*ch*). There is no constant septum across the blood channel of the filament. I believe that an endothelium is present, but can not be positive.

The surface epithelium is of columnar, ciliated cells in the frontal region (*f₁*). At its lateral edges appear gland cells in the usual position (*gle*). They are seldom seen on the inner edge of the filament in *O. virginiana*, described by Lankester (No. 9) as being as numerous here as at the side of the frontal region in *O. edulis*. The gland cells are not spherical, but much elongated in the specimens which I have examined.

As in *Pecten*, certain cells crowd together a little distance inward from the gland cells, to form the compound straining line. The ciliated lines (*lfl*) do not interlock, but touch each other and form a barrier to foreign particles which might otherwise be carried into the water tube.

The filament at the reëntering angle (*ra*) is very greatly modified. It is essentially similar, whether or not it is extended across to the opposite lamella of the gill to form the thick partitions between the water tubes. The structures showing it to be morphologically a filament are the ciliated frontal epithelium (*f₂*), the chitinous layer, here modified into two extremely large and thick rods (*ch₂*), and, in most cases, a blood channel interior to these.

The filaments next these at the angle of the fold are broadened and shortened; their chitinous layer is much enlarged and has become rod-like on either side. In fact, they offer a transitional stage between the ordinary filament and the extremely modified filament at the reëntering angle of the folds. Such a transitional filament is that represented at *tf*.

The partitions connecting the gill lamella are lined by an epithelium of cuboidal cells (*cp*), continuous with that which lines the remaining walls of the water tube. The interior of the partition consists of many longitudinal and cross muscle fibers (*mus*), and between these often lie masses of large circular cells (*pc*), which I have only seen in the partition. A considerable portion of the partition is often occupied by blood spaces (*br*).

Muscle fibers also occur in masses interior to those reëntering angle filaments not connected with a partition. They seem to be extended in the long axis of the gill.

GENERAL CONCLUSIONS IN REGARD TO THE GILLS.

1. The epithelium of the ventral edge of the gill plates in forms like *Nucula*, *Yoldia*, and *Solenomya*, and of the outer edge of the filament in all other lamellibranchs—morphologically the same thing—is specialized by the possession of "chitin," large ciliated cells, and gland cells. This of course does not apply to the deep-sea forms with a septum and no gill. This probably served in the ancestors of the group for the procuring of food as well as for aëration. However large the palps may be, they are certainly not sufficient in *Yoldia* for obtaining food. The gills here, and probably in other forms with plate gills, are extremely active in collecting food. Besides *Yoldia*, which I have observed, this function is evident in the others, from the presence of gland cells.

2. The single row of latero-frontal or straining cells, described by Peck in *Anodon*, and the two single cell rows of *Dreissena*, are not present in any of the marine forms which I have examined, with the exception of *Mytilus*. Here there is a single row of straining cells, and interiorly from these a row of many closely crowded, ciliated cells, which I have called the cells of the compound straining line. This arrangement, I believe, will be shown to be the most usual one in lamellibranch gills.

There may be one or two of these lines on each side of a filament (including the simple lines of straining cells when they occur). I have never seen more than two. In *Solenomya* and *Anomia* there seems to be no such specialization, all the cells of the outer edge of the filament being of much the same size and equally ciliated.

I believe the function of these cells to be that of preventing foreign bodies in the water from entering between the plates or filaments of the gill, by means of their large cilia, while allowing currents of water to do so.

3. Gland cells were present at the sides of the frontal region, from one to three appearing in a section on either side. The forms in which they were seen in this position were *Solenomya*, *Arca*, *Mytilus*, *Pecten*, *Anodon*, *Venus*, *Mya*, *Anomia*, and *Ostrea*. I was not certain about the matter in *Yoldia*, though I believe that I have seen them in the gill of that form; and the fact that I have seen carmine grains firmly cemented into considerable masses on the gill of the living animal confirms this belief.

In *Mytilus* they generally appeared immediately interior to the outer row of laterofrontal cells. Lankester (No. 9) has described gland cells as occurring near the inner edge of the filament in *Ostrea edulis* quite as constantly as at the sides of the frontal region. In *Ostrea virginiana* I have found them present in this region in certain specimens, but not in all. When they occur they seem to be always nearly spherical, those outward being constantly elongated.

In some instances, gland cells occur on the inner edge of the filament, most noticeably in *Anomia*, where they are extremely abundant, very much more so than in the outer edge (Fig. 85).

4. I believe that an endothelial lining of the blood cavity of the filament or plate between the chitinous layers will be demonstrated in the majority, at least, of the Lamellibranchiata. I am quite certain of having observed it in *Yoldia*, *Pecten*, *Mytilus*, *Anodon*, *Venus* and probably in *Arca* and *Ostrea*. It is especially well shown in *Pecten* and *Yoldia*.

GENERAL CONSIDERATIONS.

Changes in structure brought about by the degeneration of the foot.—It has been demonstrated that the aortæ and sinus venosus in *Anodon* possess valves or sphincters, which are supposed to be operated at different times. If the foot is to be expanded, the sphincter of the posterior aorta closes and all the blood is driven through the anterior aorta into the foot. The supply comes into the heart at this time from the mantle. During this process, the valves of the sinus venosus are also closed, and the blood is confined to the foot, which it extends.

If the animal wishes to withdraw the foot between the valves of the shell, the valves of the anterior aorta, which leads to the foot, are closed, and the sphincter of the posterior vessel extending to the mantle, as well as the valves of the sinus venosus, is opened. The muscles of the foot, together with the retractor muscles, now contract and force the blood into the sinus venosus, thence to the gills, heart, and then through the only open aorta, the posterior one, into the mantle.

As has been said, *Ostrea* has no foot, though its ancestors must have possessed one, and the mantle, being part of the mechanism used by many forms with a large foot for its protrusion and retraction, has correspondingly lost most of its great blood spaces, which were capable of holding immense quantities of blood. The mantle has become firm by the substitution of compact tissue and in *Mytilus* and other forms by the reproductive glands, and its great blood supply has been reduced to the small amount which the animal needs for the nourishment of its tissues. The posterior aorta, having once supplied the mantle mainly, is now distributed instead through the greatly developed adductor muscle.

I would point out the fact that there seems to be a correlation between the aborted or absent foot and a thick mantle with no large blood spaces; and also between a fully developed locomotor foot and a mantle consisting mainly of immense blood spaces. *Pecten* may be an exception to this rule, though I think it very probable that large blood spaces do not exist in its thin mantle. This, of course, confirms the view spoken of above, concerning the physiology of locomotion by the foot. It also shows how the condition of some of the organs in *Ostrea* and *Mytilus* has been brought about by the degeneration of the foot possessed by a locomotor ancestor.

The muscle system of Mytilus.—It will be noticed from the figures that the visceral mass is relatively small and that a large part of it is filled by the great byssus muscles. When the fixed habit was acquired and the animal fastened itself by means of the byssus, these muscles were probably developed in order to prevent a tearing of the delicate tissues when the animal was subjected to the force of the waves. This injury would easily have taken place if there had been no support to the byssus by some attachment of it to the shell. Fig. 37 shows in section a powerful muscle on either side, attached to the shell dorsally. These come together on the median line, and their combined fibers run downward and become attached to the base of the byssus organ. In Fig. 42, a side view of the main muscular system, it will be seen that any downward pull from the byssus would bring a strain to bear on all the muscles whose outer ends are attached to the shell.

Having been crowded out by the development of such a bulk of muscle tissue in the visceral mass, it was necessary for the generative gland to find another position, for a decrease in its size would lead to the extermination of the species. As we have already seen, at the same time that the development of these fixing muscles was going on there was a corresponding atrophy of the foot, which was no longer needed. But the foot, in the exercise of its locomotor function, had been dependent upon the great blood spaces of the mantle. These gradually came to be of no value and disappeared, and the generative gland pushed out into the great folds from its crowded position in the visceral mass and completely filled them.

Conclusions from a comparison of the branchial chambers.—The series of figures representing cross-sections shows at a glance the comparative size and conditions of the branchial chamber in the different forms.

The very large mantle chambers seem to be characteristic of those forms which are most active, as *Yoldia*, *Venus*, *Pecten*, and many others. In *Mytilus*, where the foot is much reduced in the adult, the surface of union of the mantle to the visceral mass is much greater, extending quite a distance down the side of the body, and so reducing the size of the branchial chamber. The foot has entirely disappeared in the oyster and the branchial chamber is much reduced, there being only room enough left for the short gills to lie between its folds (Figs. 2 and 4, Pl. LXXIX).

This footless form is a very degenerate one, and came from an active ancestor, with a fully developed locomotor foot. The comparison of the branchial chamber is of interest in this connection. In an active locomotor form, like *Venus* and others, there would be a great deal of oxidation going on in the tissues, and this would necessitate correspondingly great facilities for the aëration of the blood. Consequently we find the gills greatly developed, and a large branchial chamber is necessary in which they may be suspended. Then, also, there being a large expanse of thin-walled mantle, whose interior is made up almost entirely of great blood spaces, there is a good chance for the aëration of the blood in them, for water from the exterior bathes the mantle lobes as well as the gills, at least on their inner surfaces. It may be objected to this that when the shell is open and the foot protruded, the blood is almost entirely absent from the mantle and present in the foot. But when the animal is sufficiently buried in the sand the foot is contracted and the greater part of the blood is held by the mantle. The siphons, projecting into the water, are open, and a current is constantly running into the branchial chamber.

Suppose the power of locomotion to have been lost and the animal with its rudimentary foot to have been fastened by a byssus, as in *Mytilus*. Oxidation is lessened, for about the only tissues which sustain any wear are the large muscles which support the byssus from any shock which the animal may experience from the waves. The gills are proportionately smaller because less oxidation and less food are required; the mantle lobes lose the blood spaces and become filled with the sexual gland, and the branchial chamber is lessened in size.

The oyster has become fixed by a valve of its shell and has no byssus. The mantle blood spaces and the foot are both absent, and the form is capable of no movement whatever, excepting the closing of the shell and contraction of the mantle. The breaking down of tissues is thus reduced to a minimum and the need of the aëration of the blood is very slight indeed. So we find, as a result, an extremely small branchial chamber, admitting of small currents of water, for almost the entire space of the chamber is occupied by the gills.

It is possible that the oyster came, in its degeneration, through a form in many respects similar to *Mytilus*, from some active ancestor with a locomotor foot. Judging from the most primitive of existing lamellibranchs—*Nucula, Solenomya*, etc.—the immediate ancestors of the group probably possessed a greatly developed foot. *Ostrea*, therefore, gives evidence of having departed much further from this ancestor in its degeneration than has *Mytilus*. This latter form, in turn, is less changed by its mode of life, and, noting here and there a form on the way to the most archaic, *Venus, Arca, Nucula*, still less and less so.

And yet the foot is too variable an organ to be made an exclusive means of classification or even the chief one. *Anomia* has lost the foot on account of its fixed habit, and yet it possesses some structures which indicate a more primitive position for the form than the one commonly supposed for it. Pelseneer goes so far as to place it immediately next the primitive forms with plate gills.

But if there is one organ of lamellibranchs which is most subject to variation by secondary modification, it is the gill. While it is an important organ, it seems as if it were hardly possible to use it as a basis of classification for the whole group, as Pelseneer has done. In a group where any and all organs are so subject to secondary modification, there must be a careful comparison of many of them, instead of one or two, perhaps; and even a complete knowledge of the comparative anatomy, which we by no means possess, can not be safely used as a basis for classification without the aid of comparative embryology, which is still less known.

The phylogeny of the gills.—It is generally considered that the anatomy of those lamellibranchs which possess plate gills (*Nucula, Solenomya*, etc.) shows conclusively that the group which they form must be the most primitive one of all living lamellibranchs. More especially since the appearance of Mitsukuri's paper on these plate gills (No. 15) has this been the general opinion. Their anatomical and histological similarity to certain gasteropod gills is one of the strong points for such a belief.

But there is a very great gap between these plate gills and the strictly filamentous type, which, it seems to me, can not be explained by any facts which we now possess, either anatomical or embryological.

In his *Challenger* report on the Mollusca, Pelseneer attempts to show by a series of diagrams the phylogenetic development of the gill. Beginning with *Malletia*, with plates extending laterally, he derives *Nucula* from it, in which the outer end of the plate is turned slightly downward. He now has to interpolate an hypothetical gill in which the plates have developed ventrally for some distance, but which shows no sign of an ascending portion. The next stage which he takes in his phylogenetic development is represented by the gill of *Arca*, in which there is a fully formed ascending limb of the filament and neither limb shows anything plate-like, but both are cylindrical. But right here is the gap referred to, and it would still remain very great even if his hypothetical gill just preceding it were really known to exist.

This scheme of historical development was molded upon what little knowledge we have of the ontogeny of the gill. This knowledge, except for a few fragmentary observations, we owe to Lacaze-Duthiers (No. 7). In his study of the development of the gill of *Mytilus*, he shows that the gills appear from a ridge running horizontally along the side of the body, from which rods grow out and descend ventrally. These are separate from one another and become the filaments. After attaining to a certain length their enlarged ends fuse together and form a solid membrane. The lower edge of this membrane now grows and turns upward. There is *no mechanical bending of the filaments*, but a *growth* upward of their fused ends. After the outer edge of this membrane formed by the fused ends of the filaments has grown upward for some distance, the inner portion—that first formed—divides so as to mark out the ascending filaments as continuations of the descending ones. It is possible that this fusion of the ends of the descending filaments may not take place in the developing gills of forms like *Arca*, where in the adult the contiguous filaments are nowhere fused with one another.

Pelseneer (No. 17) now regards *Nucula* as the most primitive form. He believes *Anomia* and also the *Arcidæ* to be directly descended from forms with plate gills. As I have said, these relations may be the true ones, and yet it seems difficult to explain the structure of the gills of the former in terms of those of the latter. One might much more easily suppose, *a priori*, that the latter had developed from the filamentous type, perhaps by the degeneration of the ascending portion of its filament.

The point which I wish to make is shown in the accompanying cuts. Fig. 1 represents two gill plates of one side of the body of *Nucula*. The shaded portion represents the chitinous ventral edges. The ends of the plates here point downward and outward. Fig. 2 represents Pelseneer's hypothetical type, in which the plates are much elongated ventrally and have become more like filaments. In such a series Pelseneer could not place *Solenomya* or *Yoldia*, but had to throw them out because their plates, instead of extending their outer ends downward, extended them in quite the opposite direction.

Now come the simplest filamentous gills, those of *Arca*, Fig. 3. (I have figured the gills of *Arca pexata*.) In this the ascending filament (*afl*) is fully formed, and the difference between this condition and that of *Nucula* is seen to be very great, notwithstanding this hypothetical form.

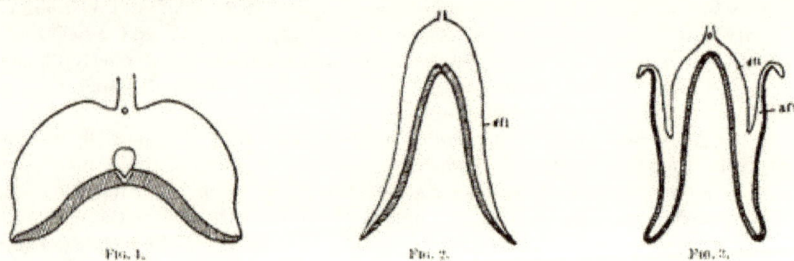

FIG. 1.   FIG. 2.   FIG. 3.

It is possible that the ascending limb of the filament is a new structure which has suddenly developed in those forms most closely connected with the forms with plate gills; if it is, however, merely a continuation outward of the descending filament, it seems as if we ought to regard the gill plate of *Nucula* as being homologous, not alone to the descending limb of a filament, as Mitsukuri has done, but to both descending and ascending limbs.

LIST OF PAPERS CITED.

1. BARROIS. Le stylet cristallin des Lamellibranches. Rev. Biol. Nord, p. III.
2. BROOKS, W. K. An Organ of Special Sense in the Lamellibranchiate genus Yoldia. Proc. Amer. Ass. Adv. Sci., 1874.
3. ———. Development of the American Oyster. Report of the Commission of Fisheries of Maryland, 1880.
4. DALL, W. H. Nature, vol. 34, p. 122.
5. ———. Preliminary Report on the Collection of Mollusca and Brachiopoda obtained in 1887-88. Proceedings of the United States National Museum, vol. XII, pp. 219 to 362.
6. GROBBEN. Die Pericardialdrüse der Lamellibranchiaten. Arb. Zool. Inst. Wien, Bd. VII.
7. LACAZE-DUTHIERS. Mémoire sur le développement des branchies des mollusques acéphales lamellibranches. Ann. Sci. Nat. Zool., ser. 4, t. II.
8. LANKESTER, E. R. Mollusca. Enc. Britt.
9. ———. On Green Oysters. Quar. Jour. Mis. Sci., vol. 26.
10. LEYDIG, FRANZ. Ueber Cyclas cornea. Müller's Archiv., 1885.
11. LOVÉN. Bidrag till Kannedomen om Utvecklingen af Mollusca Acephala Lamellibranchiata. Memoirs of the Academy of Stockholm, 1848.
12. MILNE-EDWARDS. Leçons sur la physiologie et l'anatomie comparée.
13. MITSUKURI. On the Structure and Significance of Some Aberrant Forms of Lamellibranchiate Gills. Quar. Journ. Mis. Sci., vol. XXI, 1881.
14. OSBORN, H. L. Structure and Growth of the Shell of the Oyster. Studies of the Johns Hopkins Univ., vol. II.
15. PATTEN, W. Eyes of Mollusks and Arthropods. Mitt. aus der Zool. Stat. zu Neapel, vol. VI, 1886.
16. PECK, R. H. The Minute Structure of the Gills of Lamellibranch Mollusca. Quar. Journ. Mis. Sci., 1887.
17. PELSENEER, PAUL. Contribution à l'étude des lamellibranches. Arch. de Biol., tome XI, par. II, 1891.
18. ———. Report on the Anatomy of the Deep Sea Mollusca. Zool. Challenger Expedition, Part LXXIV.
19. POSNER. Ueber den Bau der Najadenkiemen. Archiv für mikros. Anat., 1875.
20. RANKIN. Ueber das Bojanus'sche Organ des Teichmuschel. Jena, Zeit., Bd. XXIV.
21. RAWITZ, B. Der Mantelrand der Acephalen. Jena, Zeit., Bd. XXIV.
22. THIELE, J. Die Stammesverwandtschaft der Mollusken. Jena, Zeit., 1891.
23. VON JHERING. Ueber Anomia. Zeit. f. wiss. Zool., Bd. XXX.
24. ZIEGLER. Die Entwickelung von Cyclas cornea. Zeit. f. wiss. Zool., Bd. XLI.

EXPLANATION OF PLATES.

[Drawings by the author.]

PLATE LXXIX.

Ostrea virginiana—Oyster:

Fig. 1. Vertical transverse section of anterior edge of mantle (*m*), outer and inner palps (*op* and *ip*), and *br*, blood vessel.

Fig. 2. Same, through posterior end of stomach, *s*: *g*, generative mass; *l*, liver; *i*, intestine; *w*, water-tube of gill; *ol*, outer lamella of gill; *il*, inner lamella; *og*, outer gill; *ig*, inner gill; *brc*, branchial chamber; *mc*, mantle edge.

Fig. 3. Same, posterior to stomach: *epc*, epibranchial chamber; *ilc*, interlamellar connection.

Fig. 4. Same, through region of heart: *r*, rectum; *h*, heart; *p*, pericardium.

PLATE LXXX.

Ostrea virginiana—Oyster:

Fig. 5. Slightly posterior to last. The epibranchial chamber of the right side of the animal (and of the section, *epc*) extends dorsally and opens to the exterior above, between the mantle edges. The mantle is here figured as being applied to the body above, but in reality it is entirely free from it.

Fig. 6. Same, through anterior of adductor muscle, *a*.

Fig. 7. Same, through middle of adductor muscle, *a*. This figure shows the generative mass, *g*, extended nearly to the end of the rectum, *r*, dorsally, and also far beneath the adductor.

Fig. 8. Same, through posterior of adductor: *g*, generative mass at extreme posterior end of visceral mass; *cl*, cloacal chamber.

Fig. 9. Same, through cloacal chamber posteriorly; *x*, ridge running backward on inner wall of mantle, continuous with line of concrescence of gill to mantle.

Venus mercenaria—Quahog:

Fig. 10. Vertical, transverse section through mouth, *mo*, and œsophagus; *s*, anterior end of stomach, into which opens œsophagus: *l*, liver; *g*, generative mass; *pl*, palp (the anterior).

PLATE LXXXI.

Venus mercenaria—Quahog:

Fig. 11. Vertical, transverse section through stomach, *s*; *me*, mantle edge; *l*, liver; *g*, generative mass; *ar*, anterior foot-retractor; *re*, epibranchial chamber; *ig*, inner gill; *m*, mantle.

Fig. 12. Same, through junction of anterior foot-retractors, *ar*, with the foot, *f*: *ld*, liver duct.

Fig. 13. Same, through posterior end of liver; *mas*, scattered transverse muscle fibers, just beneath visceral mass. Three marked lines of vertical muscles are also shown.

Fig. 14. Same, through region of posterior end of stomach, *s*; *il*, inner lamella of inner gill; *mf*, transverse muscle fibers in the generative mass, *g*.

PLATE LXXXII.

Venus mercenaria—Quahog:

Fig. 15. Vertical, transverse section in region of anterior end of heart, which contains rectum, *r*; *p*, pericardium; *au*, auricle; *g*, generative mass; *i*, intestine; *me*, mantle edge; *ec*, epibranchial chamber; *il*, inner lamella of inner gill; *og*, outer gill; *mf*, muscle fibers in visceral mass; *f*, foot.

Fig. 16. Same, through region of ventricle of heart, *h*; *p*, extremely large pericardium; *ngl*, nonglandular part of nephridium; *gl*, glandular part of nephridium; *ol*, outer lamella of outer gill, showing foldings.

Fig. 17. Same, through region of posterior end of heart, *h*; *wn*, wall between nephridia; *pr*, posterior foot-retractors joining foot, *f*; *n*, nephridium.

PLATE LXXXII—Continued.

Venus mercenaria—Quahog—Continued.

FIG. 18. Same, in region of middle of posterior adductor, *pa*; *m*, thickened muscular mantle edge; *ws*, lower wall of branchial siphon seen posterior to branchial membrane, *brm*. (This represents the anterior surface of the section.)

FIG. 19. Same, in region of base of siphon. *as*, upper or anal siphon; *ls*, lower or branchial siphon; *c*, point where branchial siphon remains continuous with mantle; *m*, mantle continued across below siphons; *ss*, siphonal septum.

FIG. 20. Same, just posterior to last. Siphons still beneath posterior adductor, *pa*, separated from mantle, which is yet continuous beneath them.

FIG. 21. Same, posterior to last, mantle not continuous beneath siphons. Lumen of branchial siphon, *ls*, assumes oval outline.

FIG. 22. Same, through free siphons.

Mya arenaria—Long clam :

FIG. 23. Vertical, transverse section in region of anterior adductor. *a*; *mc*, thick mantle edge; *c*, cuticle from mantle edge; *m*, mantle below adductor; branchial chamber closed in by concrescence of mantle edges below; *f*, foot much flattened laterally.

PLATE LXXXIII.

Mya arenaria—Long clam :

FIG. 24. Vertical, transverse section in region of anterior end of visceral mass; *me*, mantle edge; *oe*, thick-walled œsophagus; *op*, outer palp; *ip*, inner palp; *m*, mantle; *g*, generative mass; *l*, liver.

FIG. 25. Same, in region of anterior end of stomach, *s*; *ig*, inner gill; *og*, outer gill; *re*, epibranchial chamber; *i*, intestine; *est*, crystalline style running forward on ventral side of abdomen.

FIG. 26. Same, in region of posterior end of stomach; *br*, blood vessel; *est*, crystalline style, originating from ventral wall of stomach, *s*, and running on right side of body to bottom of visceral mass. It then turns forward, and is shown in section in the preceding figure.

FIG. 27. Same, in region of heart. *h*; *n*, nephidium; *ol*, outer lamella of outer gill; *mf*, muscle fibers in generative gland.

FIG. 28. Diagram to show plane of sections through folded gills of *Mya* and *Venus*, explaining reason of folded appearance in vertical sections of the whole animal.

FIG. 29. Section just anterior to base of siphons, the anterior face of section; *r*, rectum, below plane of section, opening into continuation of epibranchial or the cloacal chamber, *cl*; *gl*, outer lamella of gill in concrescence with mantle. Gills here form complete septum between branchial and epibranchial chambers; *bs*, base of branchial siphon below plane of section; *ws*, siphonal walls; *brm*, fold at anterior end of siphonal septum, probably representing branchial membrane of *Venus*, *Mactra*, etc.; *ct*, jelly-like layer, clothing posterior end of mantle and extending out over siphons.

FIG. 30. Same, through base of siphons, right side cut a little deeper than left. The thickened mantle walls, *ws*, in the previous figures, become here the siphonal walls, *ws*, and are now much thicker; *cl*, upper end of cloaca opening below into anal siphon, *ws*; *ss*, siphonal septum.

FIG. 31. Same, through siphons at a distance from mantle; *br*, blood vessel. A similar vessel is present above the anal siphon.

Mytilus edulis—Common mussel :

FIG. 32. Transverse, vertical section through anterior end of body; *l*, liver mass; *og*, outer gill; *ig*, inner gill; *m*, thickened mantle, containing sexual gland. That portion on the median line has been drawn in greatly on killing, and thus appears in section; *ar*, anterior retractor; *pl*, palp; *aa*, anterior adductor; *me*, mantle edge.

PLATE LXXXIV.

Mytilus edulis—Common mussel:

Fig. 33. Transverse, vertical section in region of mouth, *mo*; *l*, liver; *ol*, outer lamella of outer gill; *il*, inner lamella of inner gill; *m*, thickened mantle, containing sexual gland; *pl*, anterior palp; *me*, mantle edge, joined anteriorly (*me*) down to a position almost beneath the mouth; *ar*, anterior retractor of foot; *f*, foot extending forward and flattened dorso-ventrally.

Fig. 34. Same in region of stomach, *s*; *p*, anterior end of pericardium; *l*, liver; *g*, generative mass. (The visceral mass is seen to be relatively very small.)

Fig. 35. Same, in region just anterior to base of foot; *h*, anterior end of heart; *w*, water tube between gill lamellæ; *ar*, anterior foot-retractors close together, and in the following figures are seen to run between the only retractor muscles of the foot. *fm*; *i*, intestine.

Fig. 36. Same, in region of base of foot, *f*; *fm*, posterior retractors of foot, running directly upward to become attached to shell; *ar*, anterior-foot retractors, running between posterior retractors to become attached to byssus organ.

Fig. 37. Same, in region of byssus organ and byssus. *b*; *h*, heart, containing rectum; *r* (auricles indicated at sides); *bm*, byssus muscles attached to shell and combining to support byssus organ.

Fig. 38. Same, in region of posterior end of heart, *h*.

PLATE LXXXV.

Mytilus edulis—Common mussel:

Fig. 39. Vertical, transverse section just in front of posterior adductor; *bm*, most posterior of byssus muscles; *ab*, posterior region of abdomen or visceral mass; *r*, rectum.

Fig. 40. Same, in region of anterior end of posterior adductor; *i*, intestine; *r*, rectum: *pa*, posterior adductor; *ab*, extreme posterior end of visceral mass.

Fig. 41. Same, in region of posterior end of posterior adductor.

Fig. 42. Side view of chief muscle system of *Mytilus*; *aa*, anterior adductor; *pa*, posterior adductor; *ar*, anterior foot-retractors, running backward between posterior foot-retractors, or muscles suspending the foot, *fm*, and joining the byssus organ. The muscles of the foot itself are entirely free from any connection with these. This organ, *f*, and its retractors, *fm*, may be removed from all the other muscles without anywhere injuring them; *bm*, byssus muscles; *bys*, byssus.

Pecten irradians—Scallop:

Fig. 43. Vertical, transverse section in region of mouth; *l*, liver; *mo*, large mouth opening into very wide œsophagus; *ap*, anterior palp; *fr*, part of fringe of anterior lip, the median portion of the two anterior palps.

Fig. 44. Same, in region of stomach, *s*, the walls of which are extremely uneven and folded; *i*, intestine.

PLATE LXXXVI.

Pecten irradians—Scallop:

Fig. 45. Vertical, transverse section in region of heart; *p*, large pericardium; *ao*, opening at base of pericardium into the large nephridium, *n*, below; *i*, intestine; *g*, generative mass; *a*, portion of the adductor muscle; *ven*, ventricle of heart with rectum, *r*, attached to its dorsal wall; *au*, auricle of heart running downward to connect with large vein from gills, *gl*.

Fig. 46. Same, in middle region of adductor, *da*, small, closely contracted portion of adductor with dark, striated fibers; *a*, main portion of adductor, composed of white fibers; *gm*, membrane suspending gills.

Fig. 47. Same, upper, posterior portion of adductor, showing position of greatly contracted dark portion, *da*, and mantle edges, *me*, not fused above.

Yoldia limatula:

Fig. 48. Horizontal section through inner edges of gill plates in *Yoldia* (Fig. 8), *d*); *ml*, median line; *a*, space in interior of a plate; *b*, space between plates; *mf*, muscle fibers, occasionally showing nuclei, in interior of plate; *lac*, lacunar tissue.

Mytilus edulis—Common mussel:

Fig. 49. Diagram of lamellæ of two gills; *br*, blood vessel in ends of outer lamella, continuous throughout all filaments; *ilc*, interlamellar connection; *ifc*, ciliated, interfilamentar connections.

PLATE LXXXVI—Continued.

Arca (Argina) pexata:

FIG. 50. Single gill filament. *il*, inner lamella; *ol*, outer lamella; *ifc*, interfilamentar connection.

Cardita borealis:

FIG. 51. Vertical, transverse section in region of pedal ganglion; *s*, stomach; *l*, liver; *g*, generative mass; *ig*, inner gill; *pp*, posterior palp; *me*, mantle edge; *pg*, pedal ganglion, with cerebral commissure running dorsally.

PLATE LXXXVII.

Voldia limatula:

FIG. 52. Removed from shell; right mantle fold dissected off; *aa*, anterior adductor; *pa*, posterior adductor; *l*, liver mass and generative gland, on the surface of which may be seen the loop of the intestine, *i*; *p*, large palp, extending beneath visceral mass. At its posterior end springs the appendage *ap*; *g*, gill, extending from posterior end of palp to base of siphons, *s*; *me*, mantle edge; *f*, foot; *d*, ventral disc of foot.

FIG. 53. Transverse section across siphon of *Voldia*, showing portion of wall of anal, *as*, and branchial siphons, *br*, including a portion of siphonal septum, *ss*; *c*, covering of exterior of siphons; it is more or less transparent, contains numerous nuclei, and, at places, indications of elongated cell boundaries; *trm*, regularly arranged bundles of transverse muscles in walls of both siphons and siphonal septum; *lm*, numerous bundles of longitudinal muscles (here cut transversely), alternating regularly with the transverse layers.

Ostrea virginiana—Oyster:

FIG. 54. Portion of secretory epithelium of nephridium in *Ostrea*; *exc*, excretory cells, with deeply stained bases and transparent globular free ends; *l*, lumen of excretory tube, into which break off fluid globular ends of cells. Many of these contain a small stained body.

Mactra solidissima:

FIG. 55. Crystalline style. End teased off in spiral from softer central axis, *ca*.

Pecten irradians—Scallop:

FIGS. 56 and 57. Concretions showing concentric structure, from excretory cells of nephridium.

FIG. 58. Excretory cell from nephridium; *con*, distally placed concretion (these all show a deeply stained central portion of various shapes and sizes); *vac*, numerous small vacuoles; *nuc* nucleus.

FIG. 59. Elongated cells from same, containing numerous spherical vacuoles of varying sizes.

FIG. 60. Segmenting eggs found in nephridium.

Solenomya velum:

FIG. 61. Single siphonal opening in mantle. A, walls not bent; B, walls bent to form upper, *as* and lower siphonal openings, *bs*; *t*, tentacles.

Ostrea virginiana—Oyster:

FIG. 62. Transverse section of palp of *Ostrea*, near ventral or outer edge. Inner surface thrown into ciliated ridges or folds, *f*. In this region these consist of two secondary folds, *sf*; *ct*, irregular membrane at base of folds; *fc*, fat cells.

FIG. 63. Same, midway between ventral edge and base, showing difference in character of folds, *f*.

FIG. 64. Same, at base. No secondary folds. Supporting tissue at base of folds much thicker, *ct*; *br*, blood vessel.

PLATE LXXXVIII.

Ostrea virginiana—Oyster:

FIG. 65. Striated muscle fibers in auricle of heart; *smf*, striated muscle fibers, generally, if not always, attached to thick, homogeneous, supporting membrane of wall, *sm*; *a*, protrusion of supporting membrane through epithelium of wall to exterior; *pgc*, pigment cells; *v*, vacuolated epithelial cells scattered throughout muscle fibers.

Arca (Argina) pexata:

FIG. 66. Cross-section of gill filaments; *glc*, gland cells; *lfl*, cilia of straining line; *ch*, chitin; *cj*, ciliated junction; *f*, frontal epithelium; *vs*, vascular space between walls connecting lower part of descending and ascending filaments. This filament was cut lower down than the other three represented in the figure (see Fig. 50).

PLATE LXXXVIII—Continued.

Yoldia, Venus, Pecten, Ostrea:

FIG. 67. A few spermatozoa, to show variation in appearance in different forms and in the same individual. "Tails" represented about half length. *a*, spermatozoön of *Yoldia*; *b*, of *Venus*, normal shape; *c*, of *Venus*, a very usual abnormality; *d*, of *Pecten*, normal shape; *e*, of *Ostrea*; *f*, of *Pecten*, an abnormal condition.

Pecten irradians—Scallop:

FIG. 68. Mouth fringe on edges of palps. A, bunch of fringe, seen from outer surface of lip; B, small portion of fringe of lip magnified, view of inner surface.

Yoldia limatula:

FIG. 69. Digestive tract of *Yoldia*. *m*, mouth; *s*, stomach; *i*, intestine; *pa*, posterior adductor; *r*, rectum.

Venus mercenaria—Quahog:

FIG. 70. Cross-section of single fold of gill; *ra*, reëntering angle of fold; *fil*, filaments; *mus*, muscle fibers of partition connecting lamellae of gills; *br*, blood-vessel in partition; *pbs*, great blood sinus from partition wall, sometimes almost filling half of a water tube of gill; *fbs*, blood sinus in inner edge of the filament, at the secondary reëntering angle of fold. This sometimes becomes, with the one opposite in the other lamella, as large as the partition sinus in the figure, the latter disappearing; *mus*, muscle fibers at inner edges of filaments.

PLATE LXXXIX.

Pecten irradians—Scallop:

FIG. 71. Generative gland; *or*, ovary; *t*, testis; *ep*, ciliated epithelium on surface of visceral mass; *glc*, gland cells; *bm*, basement membrane; *ct*, tissue of irregular cells beneath epithelium; *fep*, follicular epithelium; *d*, ciliated ducts, the one in the testis containing spermatozoa, and on its walls a gland cell is shown; *br*, blood vessel.

Ostrea virginiana—Oyster:

FIG. 72. Hermaphrodite gland; *ep*, ciliated epithelium of surface of visceral mass; *glc*, gland cell; *fep*, follicular epithelium, apparently giving rise only to ova; *or*, ovum; some are free, some attached to walls of follicle; *sper*, apparently ripe spermatozoa completely filling lumen of follicles; spermatozoa and ova occupy *same* follicles.

Venericardia borealis:

FIG. 73. Section horizontally across byssus organ; *fd*, folded secreting surface; *bs*, byssus secretion in fold; *bm*, muscles of byssus cut transversely; *c*, large clear cells near inner edges of folds.

FIG. 74. Excretory epithelium of one of folds of same, more magnified; *ec*, columnar cells over the distal ends of which is a striated layer probably of the byssus secretion, but appearing much like cilia; *bs*, evident byssus secretion; *lc*, transparent cells at base of fold, without nuclei, outer edges indistinct and striated; *br*, blood vessel with endothelial lining.

PLATE XC.

Pecten irradians—Scallop:

FIG. 75. Section across mouth fringe; *ep*, rod-like epithelial cells bearing long cilia; *bm*, basement membrane; *glc*, large, elongated gland cells, the nuclei of some being visible; *br*, blood vessel with endothelial lining.

Ostrea virginiana—Oyster:

FIG. 76. Transverse section of gill; 1, 2, 3, 4, folds of lamella between two lamellar partitions; f_1, ciliated frontal epithelium of filament of fold; f_2, ciliated epithelium of frontal region of modified filament, at reëntering angle of fold; ch_1, chitinous rods in filament of fold; ch_2, chitinous rods of modified filament; *ra*, reëntering angle of fold; *lfl*, cilia of cells of straining line; *glc*, gland cells; *mus*, muscle fibers; *pc*, spherical cells in partition connecting gill lamellæ; *wt*, water tube; *bc*, blood corpuscles; *br*, blood vessel of partition; *ep*, epithelium of partition; *tf*, filament, transitional between filament of fold and filament of reëntering angle.

PLATE XCI.

Solenomya velum :

FIG. 77. Cross-section of two gill plates; *gle*, gland cells; *cc*, ciliated, columnar cells of thickened ventral edge of plate; *ch*, chitinous rods, lined interiorly by endothelium; *w*. cells of interior walls of plate.

Yoldia limatula :

FIG. 78. Diagrammatic view of ventral edges of gill plates; *p*, ventral edges of plates: *gr*, ventral groove; *cj*, ciliated junction between plates.

FIG. 79. Vertical section across gill plate; *c*, frontal epithelium bearing very large cilia; *r*, ridge of cells forming first compound straining line; *t*, cells forming second compound straining line; *ch*, chitin, inclosing *b*, a blood-channel lined by endothelial cells; *s*, blood space of interior of plate, bridged across by numerous branching lacunar cells, *lac*; *f*, folds in wall of plate; *d*, blood space at dorsal edge of plate.

FIG. 80. Cross-section of dorsal ends of plates of gill; *lac*, numerous lacunar cells; *ep*, epithelium on surface of supporting membrane of gills; *m*, muscles of same.

FIG. 81. Diagram of plates of gills showing planes of sections; *a*, plane of Fig. 79; *b*. of Fig. 82; *c*, of Fig. 80; *d*, of Fig. 48, Pl. LXXXVI.

FIG. 82. Sections of plates of gill, *Yoldia*, in plane *f*, Fig. 81, showing lines of ciliated junction, *cj*; *c*, lateral edge of plates; *s*, space between edges of plates where ciliation is absent; *lac*, lacunar tissue; *ep*, epithelium of supporting membrane of gills; *m*, muscles of same.

PLATE XCII.

Pecten irradians—Scallop :

FIG. 83. Cross-section of single filament, gill of *Pecten*; *f*, ciliated frontal epithelium; *gle*, gland cells; *pg*, pigment in region of gland cells; *lfl*, cells of compound straining line; *ch*, chitin; *p*, septum of filament showing nucleus.

FIG. 84. Diagram of portions of three filaments of gill, to show nature of ciliated spurs, *cs*, which form means of interfilamentar union; *fil*, filament.

Anomia simplex :

FIG. 85. Inner ends of filaments of *Anomia*, to show gland cells.

Pecten irradians—Scallop :

FIG. 86. Cross-section of gill of *Pecten*; *ra*, reëntering angle of gill folds; *sa*, filament at salient angle of gill fold; *ch*, chitin of modified filament of reëntering angle; *lfl*, straining line of modified filament; *bc*, blood space of modified filament; *w*, walls of modified filament, forming interlamellar union; *lac*, lacunar tissues; *cs*, ciliated spurs connecting filaments.

PLATE XCIII.

Mytilus edulis—Common mussel :

FIG. 87. Siphonal region of mantle in *Mytilus* seen from behind; *co*, cloacal opening; *brm*, branchial membrane; *me*, mantle edge.

FIG. 88. Same, seen from in front; *co*, cloacal opening; *og*, outer gill; *brm*, branchial membrane; *m*, mantle.

Ostrea virginiana—Oyster :

FIG. 89. Posterior region of mantle in *Ostrea*, seen from behind; *r*, rectum, opening into upper end of cloacal chamber; *brp*, concrescence of mantle edges above gills; *ig*, inner gill; *og*, outer gill; *me*, mantle edge.

Venus mercenaria—Quahog :

FIG. 90. Posterior region of mantle in *Venus*, view of base of siphons; *og*, outer gill; *bm*, branchial membrane; *me*, mantle edge.

Solenomya velum :

FIG. 91. Gills of one side of body in *Solenomya; m*, supporting membrane: *gr*, groove on median line opposite supporting membrane.

Yoldia limatula :

FIG. 92. Gills of one side of body in *Yoldia ; m*, supporting membrane; *gr*, groove on midline of ventral surface.

PLATE XCIII—Continued.

Venus mercenaria—Quahog:

    FIG. 93. Longitudinal vertical section through posterior mantle region, *Venus*; *pa*, posterior adductor; *r*, rectum; *cl*, cloacal region; *epc*, epibranchial chamber; *ig*, inner gill; *brm*, branchial membrane; *brc*, branchial chamber; *me*, mantle edge.

Mya arenaria—Long clam:

    FIG. 94. Longitudinal, vertical section of posterior mantle region in *Mya*; *pa*, posterior adductor; *r*, rectum; *cs*, cloacal siphon; *bs*, branchial siphon; *brm*, fold in position of branchial membrane of *Venus*; *ig*, inner gill; *brc*, branchial chamber; *me*, mantle edge; *sf*, folds in wall of partly contracted siphon.

Mytilus edulis—Common mussel:

    FIG. 95. Anterior end of body of *Mytilus*, cut off by vertical transverse section just behind mouth, to show position of palps; *ar*, anterior retractor muscles; *ip*, inner palp; *op*, outer palp; *ig*, inner gill; *m*, mantle; *mo*, mouth.

PLATE XCIV.

Venus mercenaria—Quahog:

    FIG. 96. *Venus*, life size, right valve of shell and mantle fold being removed; *aa*, anterior adductor muscle; *pa*, posterior adductor; *afr*, anterior foot-retractor muscle; *pfr*, posterior foot-retractor; *lg*, ligament of shell; *hg*, hinge of shell; *epc*, epibranchial chamber; *h*, heart seen beneath; *pl*, anterior palp; *f*, foot; *me*, mantle edge; *ig*, inner gill; *su*, portion of mantle remaining on sides of bases of siphons.

Ostrea virginiana—Oyster:

    FIG. 97. *Ostrea*, life size, right valve of shell and mantle fold being removed; *lg*, ligament of shell; *ip*, inner palp; *ig*, inner gill; *n*, nephridium under the adductor muscle; *me*, mantle edge; *c*, cloacal region; *r*, rectum; *a*, adductor muscle; *h*, heart seen beneath the pericardium.

PLATE LXXIX.

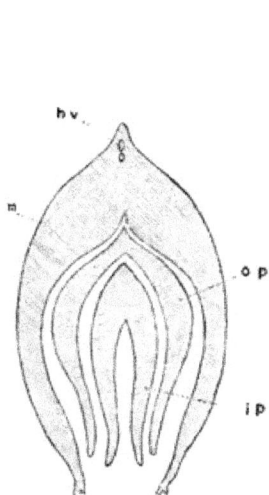

Fig. 1

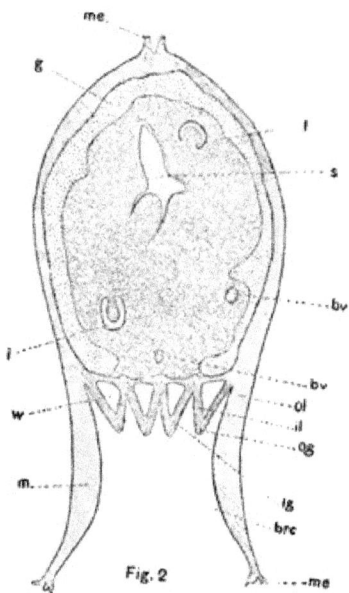

Fig. 2

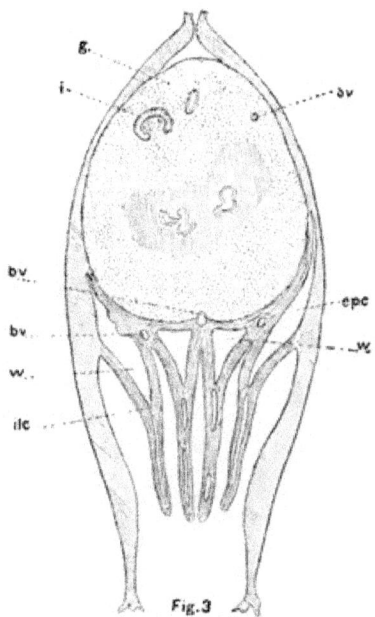

Fig. 3

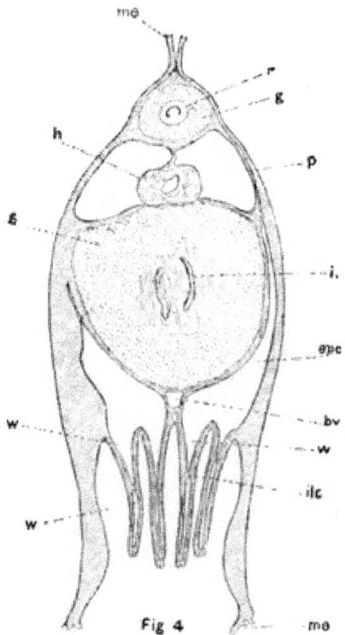

Fig 4

Kellogg del.

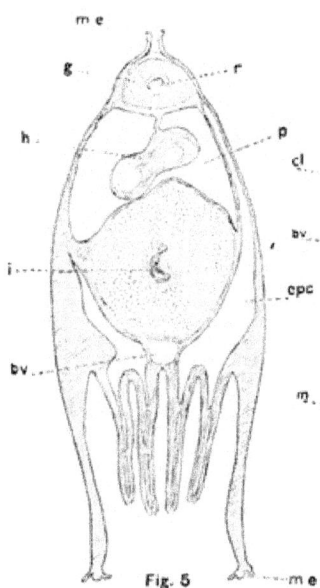

Fig. 5

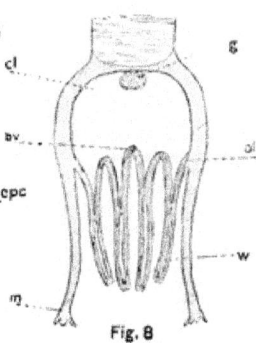

Fig. 8

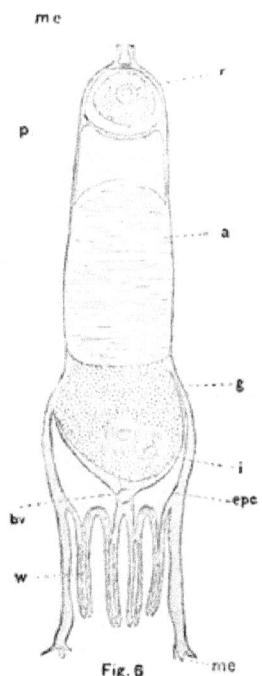

Fig. 6

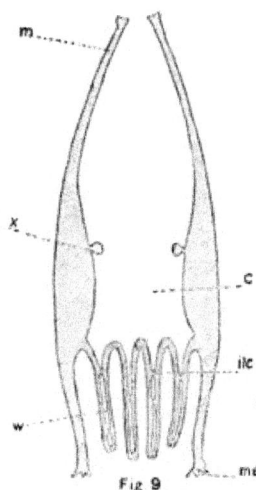

Fig 9

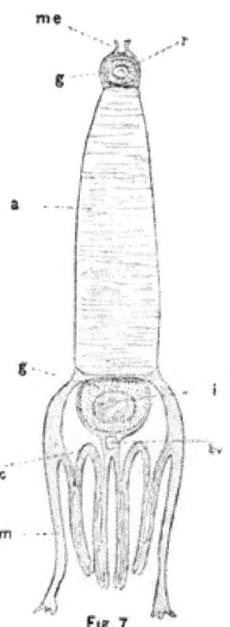

Fig. 7

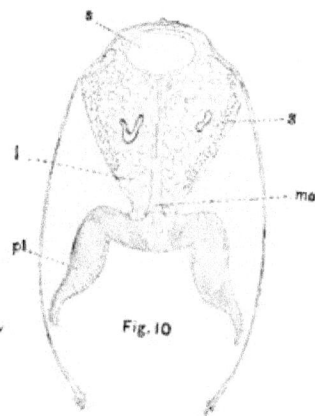

Fig. 10

Kellogg del.

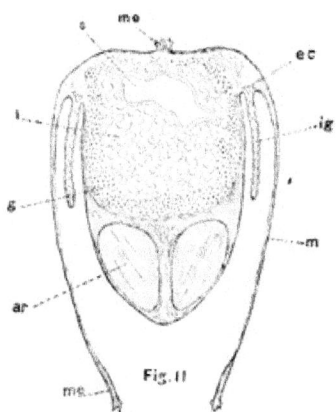

Fig. 11

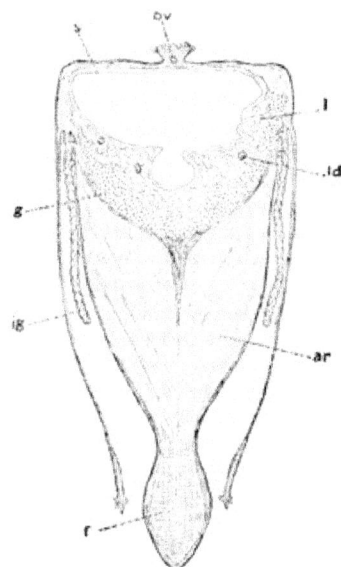

Fig. 12

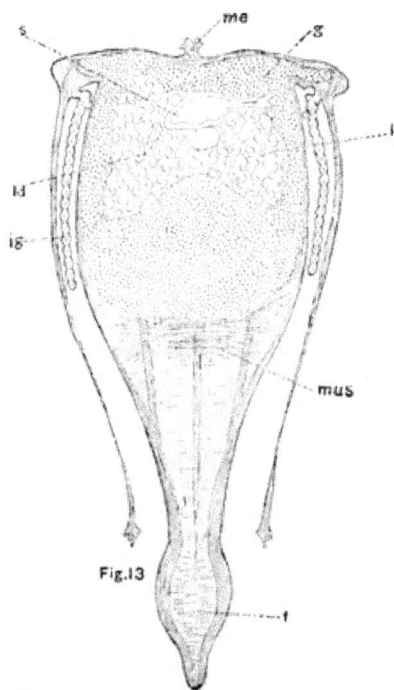

Fig. 13

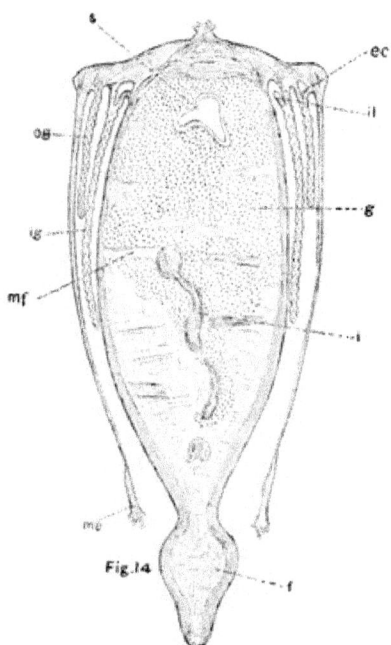

Fig. 14

Kellogg del.

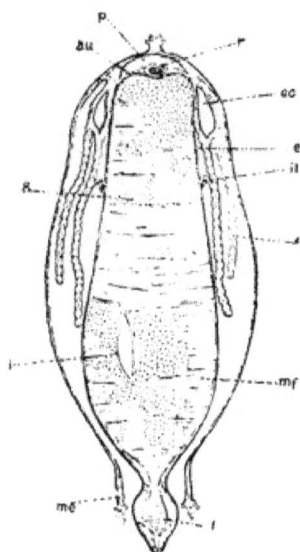

Fig. 15

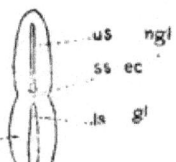

Fig. 22

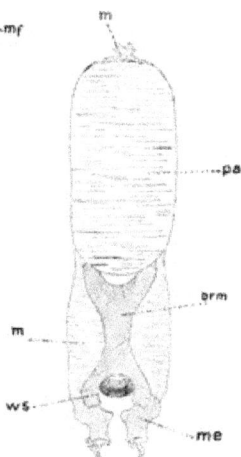

Fig. 18

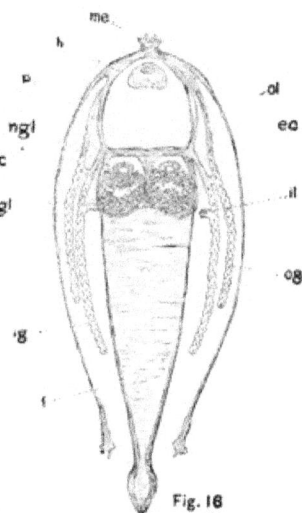

Fig. 16

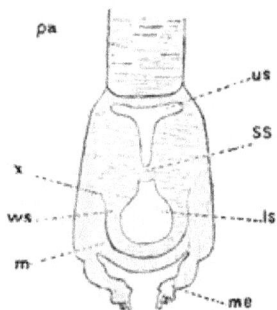

Fig. 19

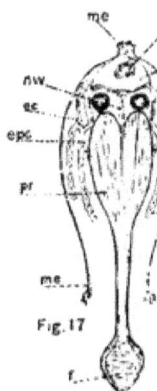

Fig. 17

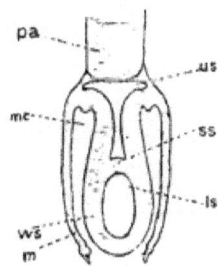

Fig. 21

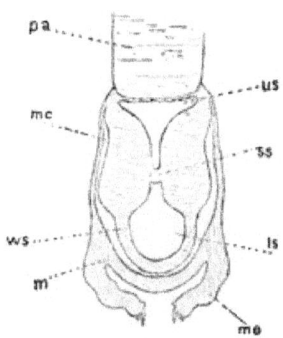

Fig. 20

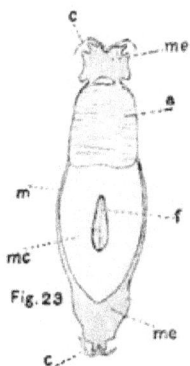

Fig. 23

Kellogg del.

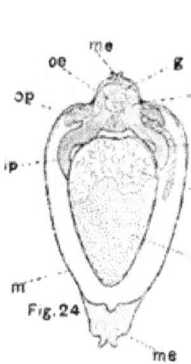

Fig. 24

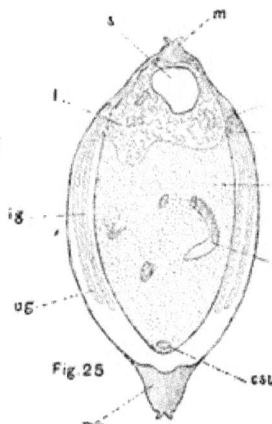

Fig. 25

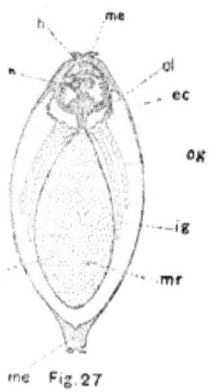

Fig. 27

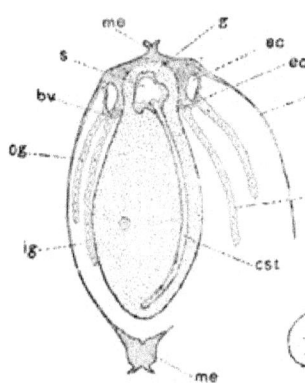

Fig. 26

Fig. 28

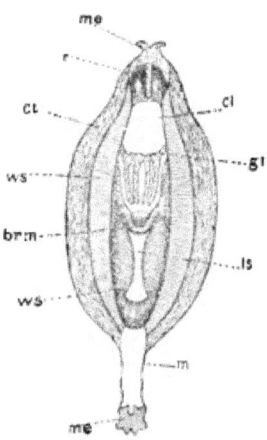

Fig. 29

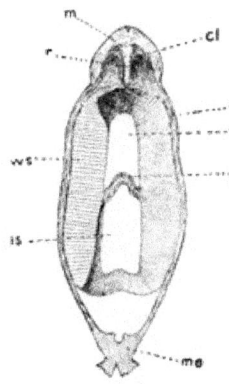

Fig. 30

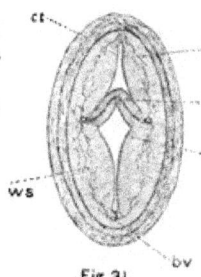

Fig. 31

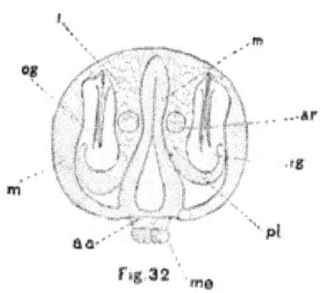

Fig. 32

Kellogg del.

Bull. U. S. F. C. 1899. Lamellibranchiate Mollusks. (To face page 42.)

PLATE LXXXIV.

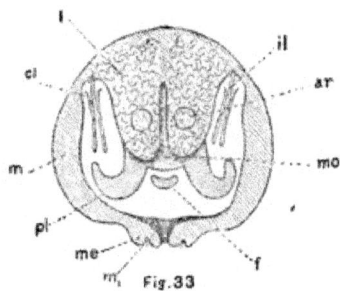

Fig. 33

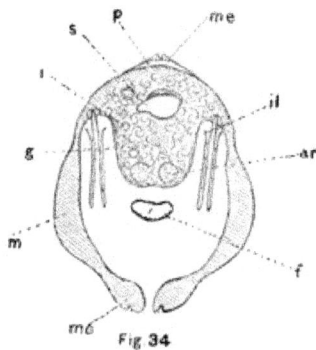

Fig. 34

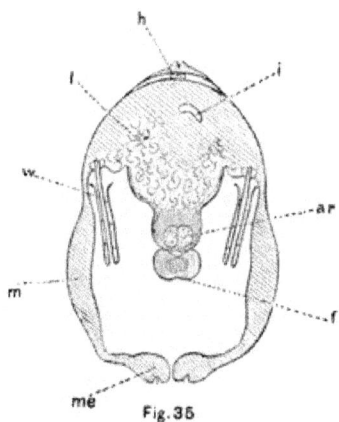

Fig. 35

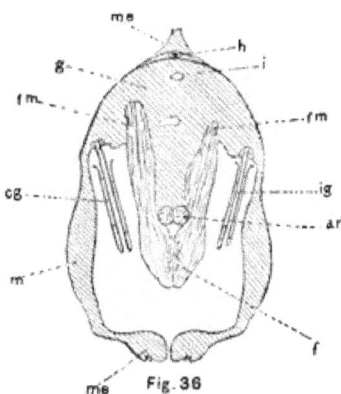

Fig. 36

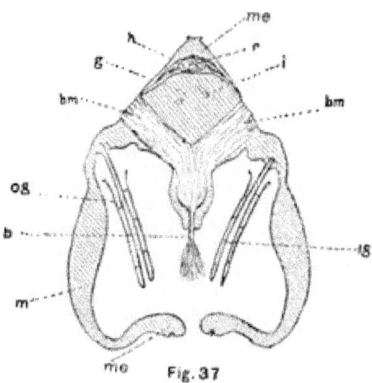

Fig. 37

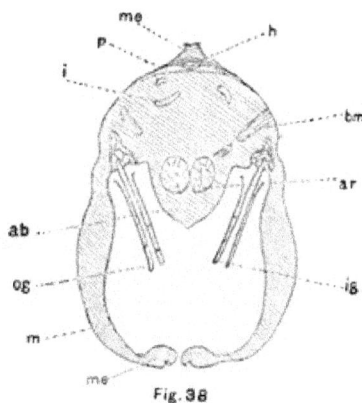

Fig. 38

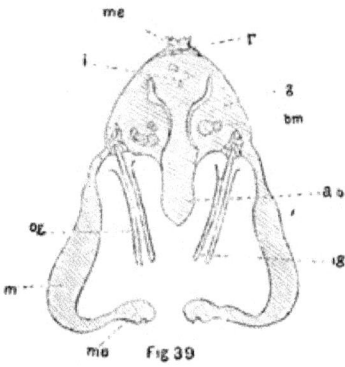

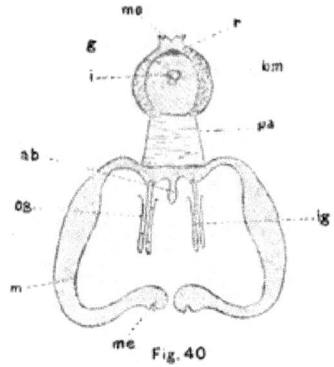

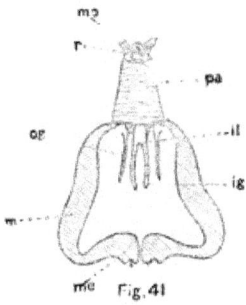

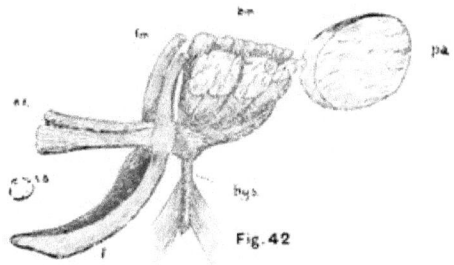

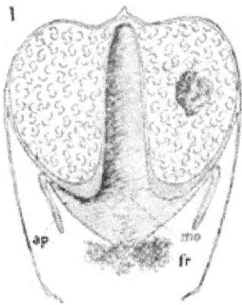

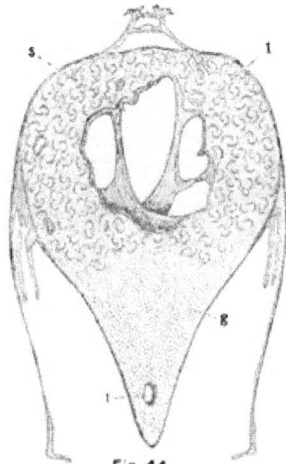

me
r
i
g
bm
og
ab
m
mo   Fig 39

me
g
r
i
bm
pa
ab
og
lg
m
me   Fig. 40

m2
r
pa
og
il
ig
m
me   Fig. 41

fm
bm
pa
st
hys
l
Fig. 42

l
ap
mo
fr
Fig 43

s
l
g
l
Fig. 44

PLATE LXXXVI.

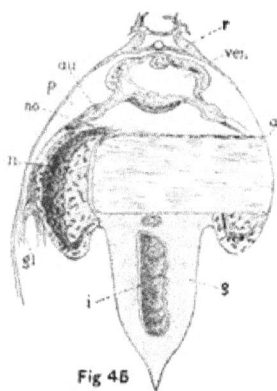

Fig 45

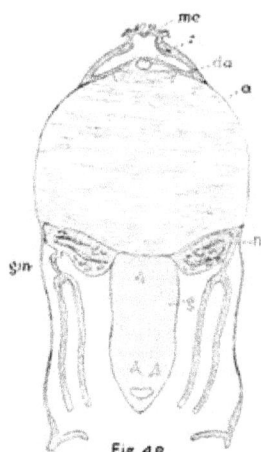

Fig. 46

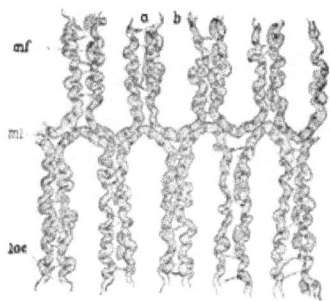

Fig. 48

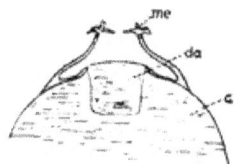

Fig. 47

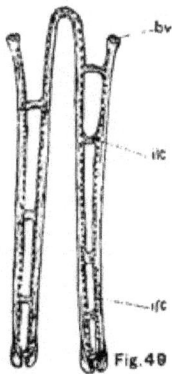

Fig. 49

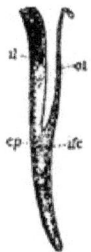

Fig. 50

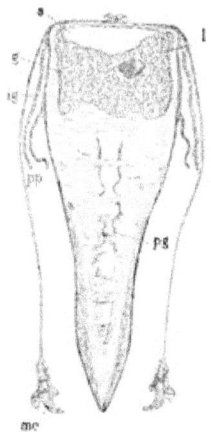

Fig 51

Kellogg del.

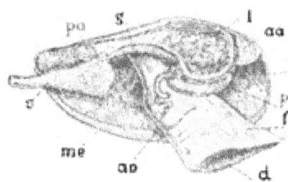

Fig. 52

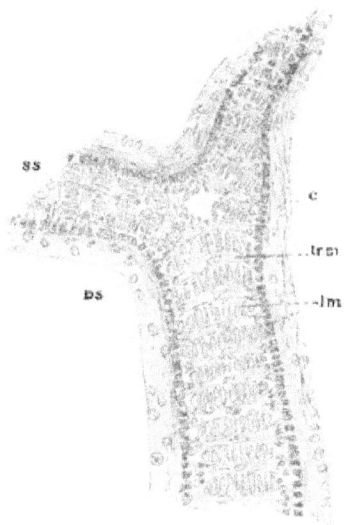

Fig. 53

Fig. 54

Fig. 55

Fig. 56 Fig. 57

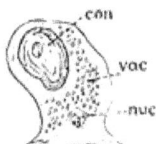

Fig. 58

Fig. 59

Fig. 60

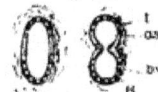

Fig. 61

Fig. 62

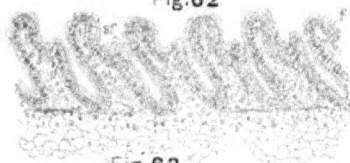

Fig. 63

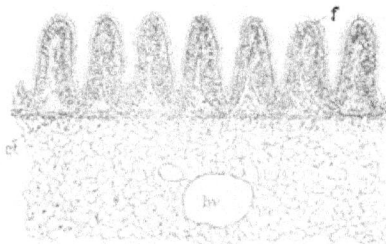

Fig. 64

Kellogg del.

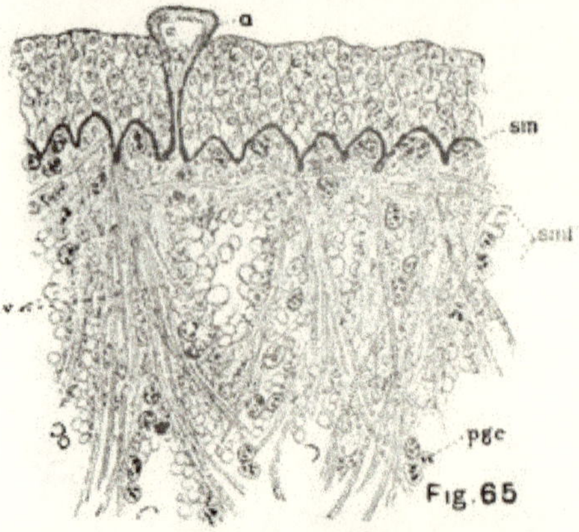

Bull. U. S. F. C. 1890. Lamellibranchiate Mollusks. (To face page 436.)

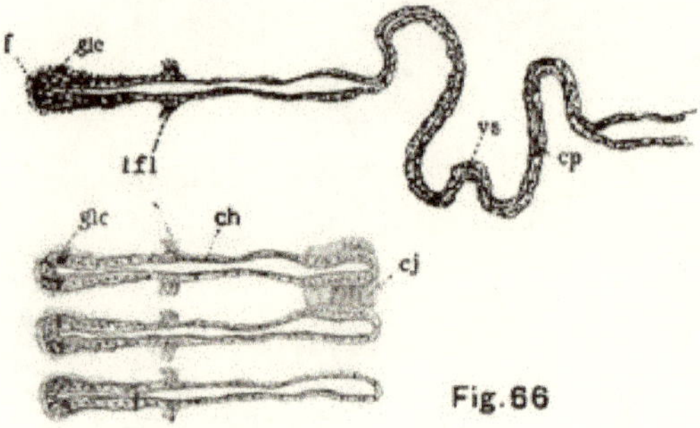

PLATE LXXXVIII.

Fig. 65

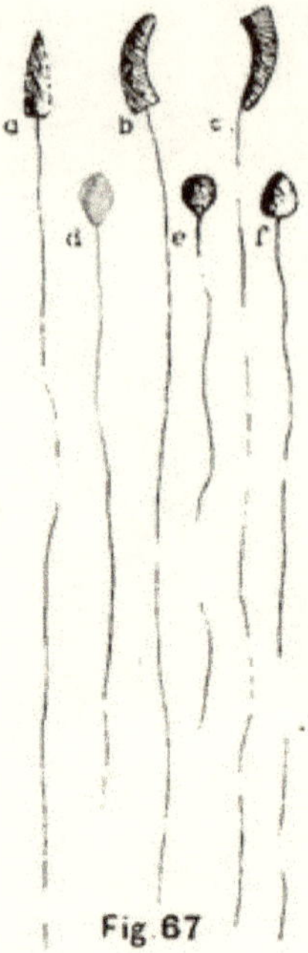

Fig. 66

Fig. 67

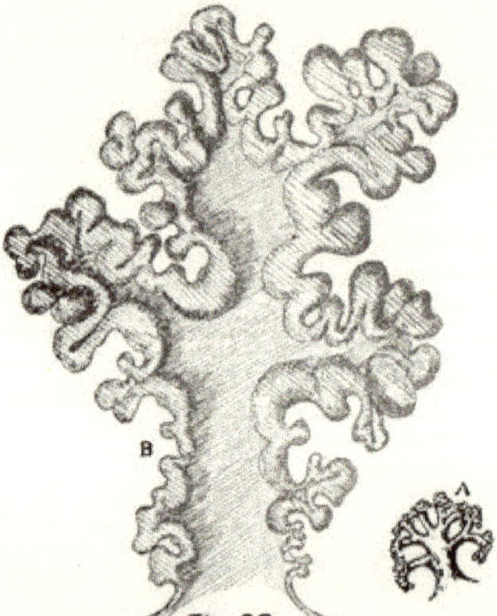

Fig. 68

Fig. 69

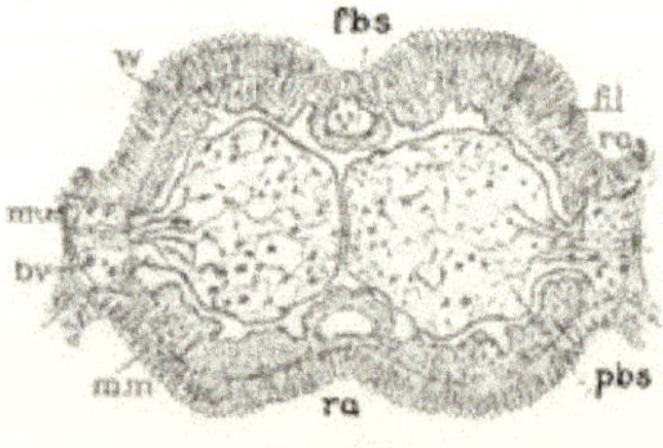

Fig. 70

Kellogg del.

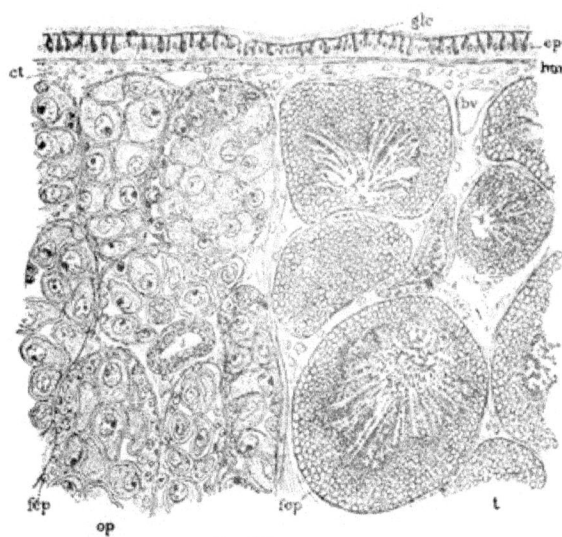

Fig.71

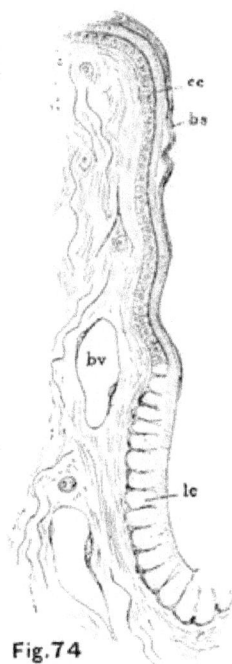

Fig.74

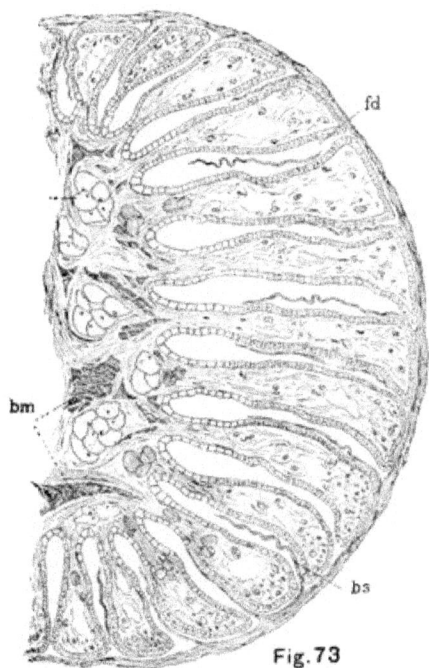

Fig.73

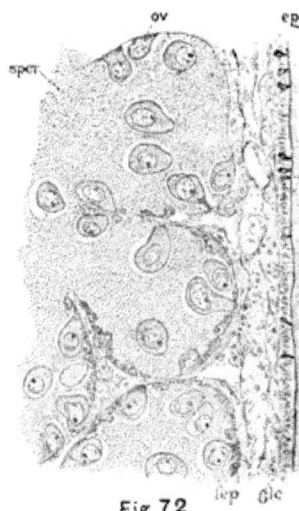

Fig.72

Kellogg del.

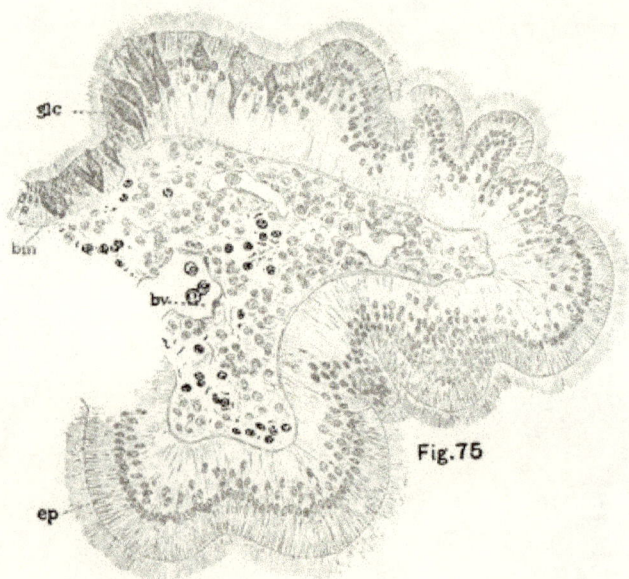

Fig.75

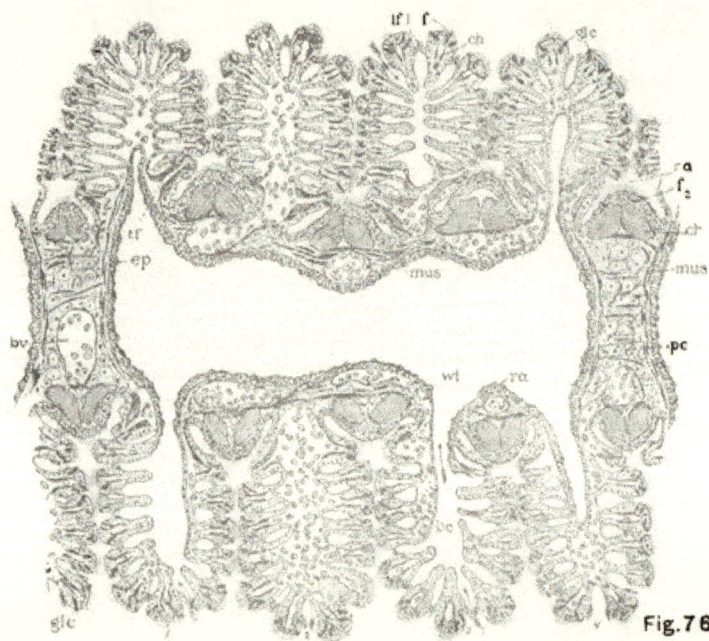

Fig.76

Kellogg del.

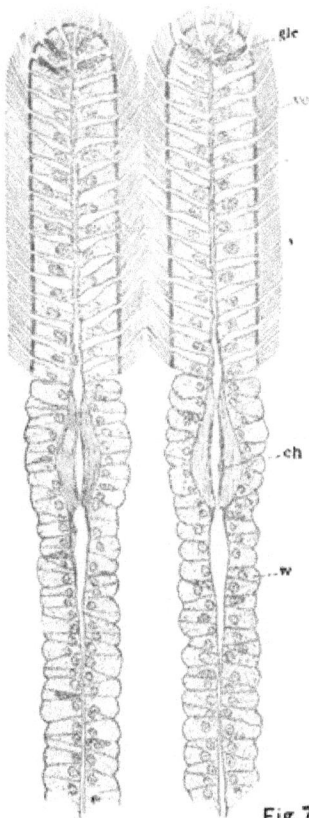

Fig.77

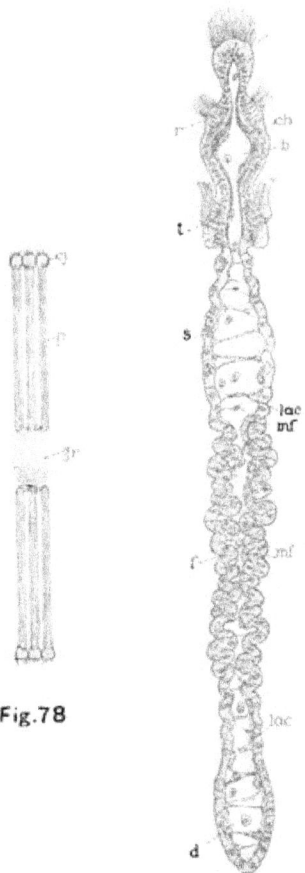

Fig.78

Fig.79

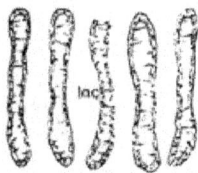

Fig.80

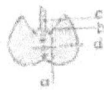

Fig.81

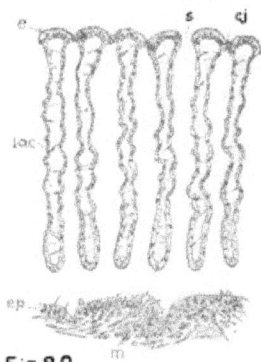

Fig.82

Kellogg del.

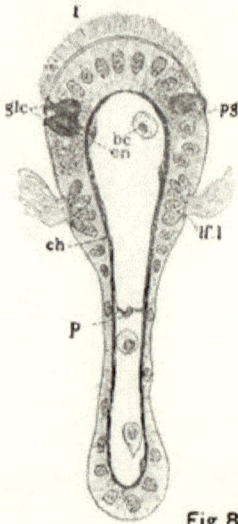

Fig. 83

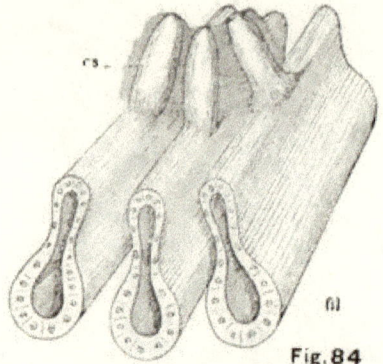

Fig. 84

Fig. 85

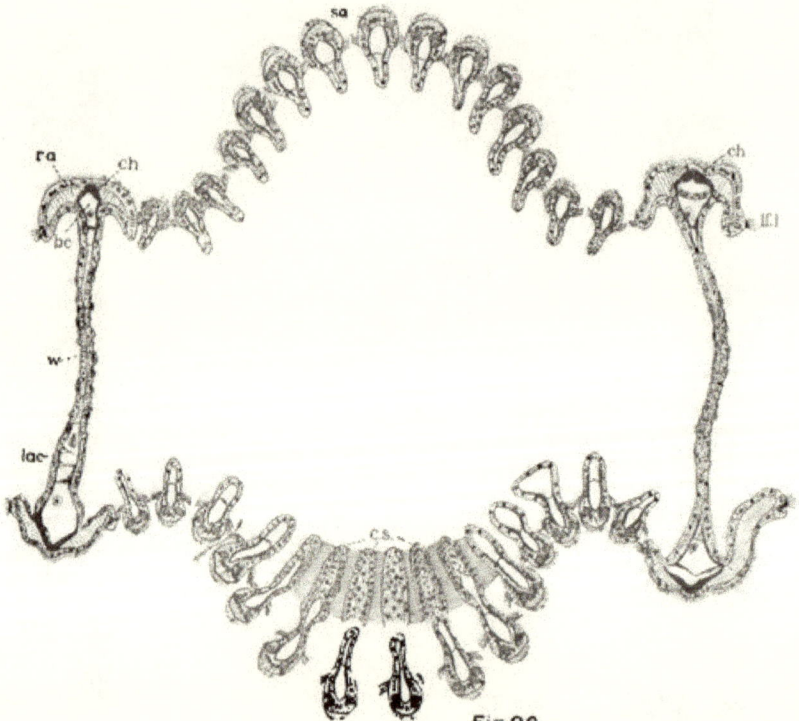

Fig. 86

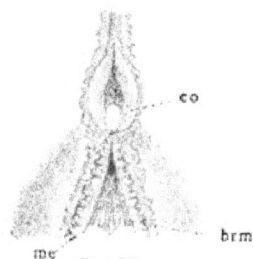

Fig.87

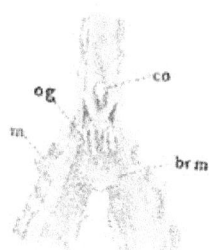

Fig 88

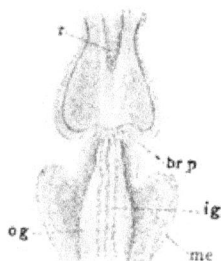

Fig.89

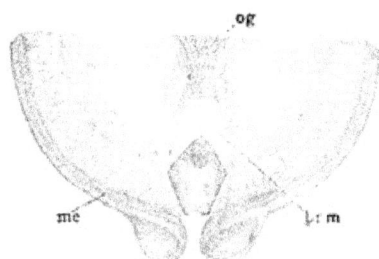

Fig.90

Fig.92

Fig.91

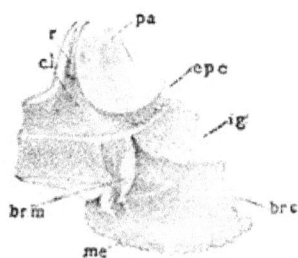

Fig.93

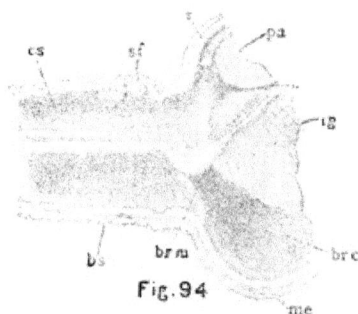

Fig.94

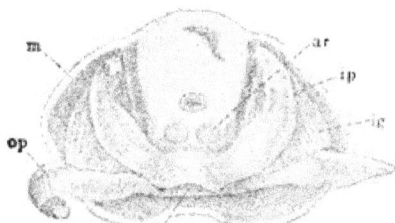

Fig.95

Kellogg del.

Bull. U. S. F. C. 1890. Lamelibranchiate Molluscs. (To face page 436.)

PLATE XCIV.

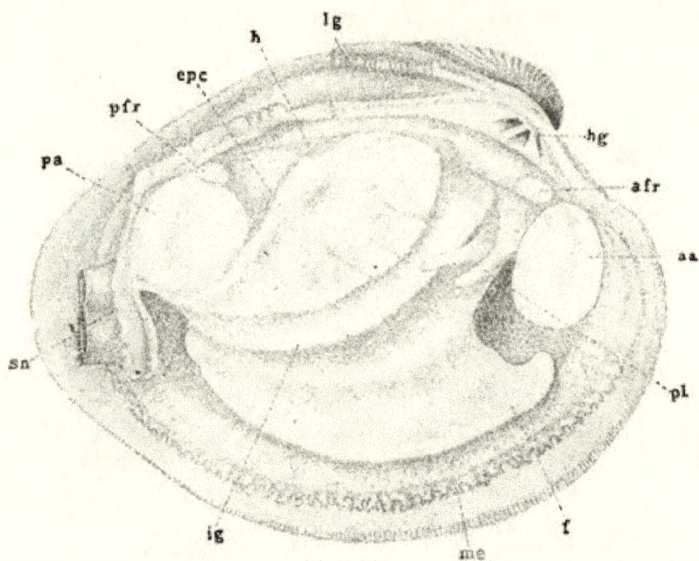

Fig. 96

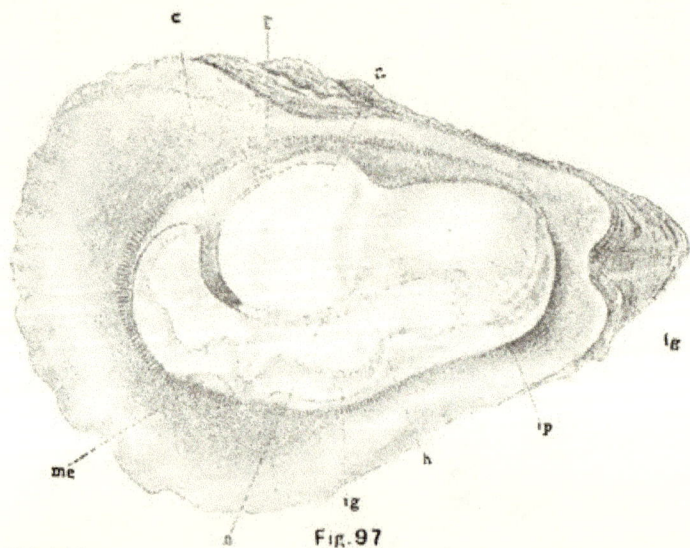

Kellogg del.

Fig. 97